Vesuvius

Vesuvius

A biography

Alwyn Scarth

TERRA

First published in 2009 by Terra Publishing

Terra Publishing
PO Box 315, Harpenden, Hertfordshire AL5 2ZD, England
Telephone: +44 (0)1582 762413
Fax: +44 (0)870 055 8105
Website: www.terrapublishing.net

ISBN-13: 978-1-903544-25-9 hardback

18 17 16 15 14 13 12 11 10 09
11 10 9 8 7 6 5 4 3 2 1

British Library Cataloguing-in-Publication Data
A CIP record for this book is available from the British Library

Library of Congress Cataloging-in-Publication Data are available

Typeset in Bembo and Helvetica
Printed and bound by MPG-Biddles Limited, King's Lynn, England

Contents

v

Preface

Vesuvius is the most famous and one of the most violent volcanoes in the world, and Naples and the province of Campania around it have the reputation of housing the most impassioned population in Europe. This biography of Vesuvius evokes the intimate relationship between the volcano and people, which has been recorded in unrivalled detail for more than two thousand years. The Campanians have never been able to remember with serenity, nor to forget with impunity, that they live in a volcanic land that has witnessed some of Europe's most powerful and lethal eruptions.

The story of Vesuvius fascinates by its rich geological or geographical history, which has been told by Earth scientists who have often been among the world's greatest experts on volcanic behaviour. But the Earth sciences are only part of the story. The other side of the narrative recounts the changing social, religious, and intellectual impact that the volcano has always had upon the population. Many vivid and fascinating eye-witness accounts have followed Pliny the Younger's description of the eruption in AD 79 and, ever since, religious beliefs, prejudice, education and fear have stimulated a whole range of human reactions to every volcanic crisis and the devastation and mourning that ensued. This study is based on the latest academic research, but also on a prudent appraisal of contemporary accounts and, wherever possible, I have based the story on eye-witness descriptions by the participants. But they have also to be examined with critical care to discover the grains of truth among the chaff of fantasy and inaccuracies that have distorted many older versions of the Vesuvian story.

Until very recently, volcanic eruptions came as bolts from the blue. Now, however, scientific experts are fast developing techniques for predicting the behaviour of volcanoes. Vesuvius is well worth the closest scrutiny, because it may be approaching its most violent outburst since 1631. Yet, one of the major current environmental and social problems in Campania is to convince the population that the next eruption really will put them in the gravest danger. Indeed, Vesuvius is not the only volcanic threat in the district. In the

Campi Flegrei, west of Naples, two periods of earthquakes have caused panic and damage within the past 50 years, although the feared eruption did not take place.

Many sources among the vast range of Vesuvian studies have been assembled in the extensive bibliography. In order to present the narrative with the vitality that the subject deserves, direct references have not been included in the text, although the most important authorities have been noted at the end of each chapter. Similarly too, the few wholly unavoidable technical terms have been explained in the glossary.

This biography draws together strands of enquiry from archaeology, the classics, the Earth sciences, history, literature, planning, politics and religion. I wrote it for all those who would welcome a thorough study of the changing relationships between Europe's most violent volcano and the people living around it.

Alwyn Scarth, Paris, June 2009

Acknowledgements

The author and publishers would like to thank the following for granting permission to use material in this book: Michael Sheridan and the US National Academy of Sciences for the photographs of some results of the Avellino eruption; and to the UK Government Art Collection for the engraving of Sir William Hamilton.

I am extremely grateful to Harry Hine, Christian Morea, and Shirley and Knud Larsen for their invaluable assistance with translations from Latin, Italian and Danish; to Derek Hopper and Tom Scotland, then of 614 Pathfinder Squadron RAF, for their photographs of the eruption in 1944; to David Alexander for criticisms of an earlier draft of this manuscript; to Rosalba and Lucia Lepore for offering me much enlightenment about the Neapolitans; and to Roger Jones for all his help and encouragement during this project.

Finally, I should like to thank Catherine Lagoutte, Pat Michie, Anthony Newton, Morag Niven, Brian Storey and David Wallace for their invaluable help, and especially Jean-Louis Renaud for improving and clarifying this manuscript.

Alas, any errors are my own.

Chapter 1

Introduction

Place the most violent volcano in Europe in the midst of one of the most volatile populations on the continent and sparks are bound to fly.

Vesuvius is the most dangerous volcano in Europe when it erupts, and the crowning glory of Campania even when it is dormant. The volcano mirrors the fascinating, effervescent, vibrant and sometimes disturbing city of Naples that faces it across a bay of legendary beauty. It is as if this splendid and paramount couple have nourished each other for centuries. Vesuvius holds the stage: a talisman for the Campanians; a manifest threat to their livelihoods and to their very lives; it destroyed Pompeii and Herculaneum, and preserved them for posterity; and it is the spirit presiding over all the contradictions in the region and its inhabitants. The volcano is all the more dangerous because it rises in the midst of a populous area that has owed its rural wealth to fertile soils weathered from the ash and lavas of previous destructive eruptions. Vesuvius plays a dual role: provider and exterminator; preserver and destroyer; guardian and enemy; tourist attraction and killer; it is a volcano often to be admired, but always obeyed; and it is a volcano with a benign and beautiful appearance that masks a ferocious temper, giving substance to the ambiguity in the famous dictum "See Naples and die".

Vesuvius has a personality that centuries of scrutiny have brought out in all its intimate detail. Some of the volcano's fascinating character and volatile behaviour seem to have rubbed off onto the people living around it. Observer and observed have developed a symbiotic relationship that shows no signs of diminishing. The Campanians are the victims; and Vesuvius is the aggressor that they fear and admire. Vesuvius puts on a show for the Campanians and they, in turn, put on a show for Vesuvius whenever the volcano springs back to life. Their interrelationship is so close that the people often resent interference from outsiders, even from those who are trying to protect them from the next eruption.

1

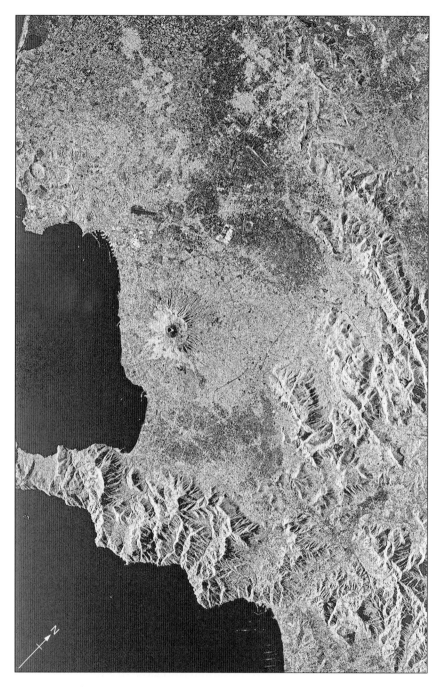

A satellite image of Campania, with Vesuvius rising in isolation in the central lowland (courtesy of NASA: image PIA01780).

Nowhere has this interrelationship been better displayed than on the flanks of Vesuvius itself. Eruptions have destroyed settlements here time and again, but the villagers have refused to be cowed, and they have rebuilt their homes almost as soon as the ash and lava have cooled. In recent decades, the bolder or more foolhardy among them have even built their homes almost to the very foot of the great cone – as if they were challenging the volcano to do its worst. This is an affront that Vesuvius will not ignore, and its retribution threatens to be ferocious.

Ancient settlements

The history of Vesuvius has always been intimately linked with the people of Naples and the towns that form a necklace around its lower flanks. Naples (originally named Neapolis, the "new city") was founded by Greek colonists perhaps as early as 800 BC. In fact, it was probably an offshoot of an even older Greek settlement that was first called Parthenope and later Paleopolis (the "old city"). Naples flourished and was eventually conquered by the Romans in 326 BC. It soon became, and long remained, the centre of one of the most delightful regions of the Roman world, a region that thoroughly deserved to be called *Campania felix* ("happy Campania"). Here flourished market towns such as Pompeii and Capua, resorts such as Herculaneum, Baiae and Oplontis, and ports such as Puteoli (Pozzuoli). Campania suffered the Barbarian invasions during the fifth and sixth centuries, and declined along with all the Roman Empire until powerful rulers established more permanent governments in the area.

Foreign rule

Except for a few months here and there, different foreign dynasties dominated and exploited Campania for a thousand years. The Norman French took over Campania in about 1040. German Hohenstaufens followed in 1190. French Angevins succeeded them in 1266. In 1442, Spanish Aragonese and then Spanish Hapsburgs began 250 years of domination through powerful viceroys. Austrian Hapsburgs started a discontinuity in the long period of Spanish power in 1707, only for the Spanish Bourbons to replace them as kings of the Two Sicilies in 1734. The Napoleonic French then ruled for nearly a decade from 1806, but the Spanish Bourbons regained control after the Battle of Waterloo in 1815. Among these rulers, a German, the Emperor Frederic, one of the Spanish viceroys, Pedro de Toledo, and a Spanish king, Charles of Bourbon, rose above their generally undistinguished rivals. Few were the rulers who actively sought to bring benefits to Campania; rather was it seen as a place to plunder like a colonial economy. Most of the Spanish viceroys, for instance, usually had to milk their charge for only a few years before they could return to Spain with galleon-loads of booty and millions of ducats. One of the greediest of them all was the Count of Monterrey, the viceroy who processed so piously during the great eruption of Vesuvius in 1631 (see Ch. 7).

Authority questioned

Then, in 1861, Campania joined the newly united Italy, and a new era seemed to have dawned. But it was a clouded dawn. The government in Rome seemed no better than the foreigners who had preceded them. The Campanians, and the Neapolitans in particular, had always used their considerable ingenuity, and the inspiration of the saints, to think for themselves. In this, their ancient Greek heritage came once more to the fore. They now retained their well developed scorn of far-away authority and, in no time at all, a new and notorious parallel government also began to hold sway. These attitudes still persist and, for many Campanians, the apparently distant prospects of another eruption of Vesuvius are far less threatening than the ever-present unemployment, corruption, crime and poverty.

The downside is that such attitudes now constitute major handicaps in making contingency plans to combat the next violent eruption of Vesuvius. This is not merely a matter of academic sociology: many thousands of people will die if these plans fail. Naples alone now has a population of over 2 million, while a further million reside around the lower flanks of the volcano.

Volcanoes began to erupt in Campania some 300 000 years ago and, for many thousands of years, two large cones stood side by side on the shores of the Bay of Naples. Then, two huge eruptions destroyed the western volcano and left behind the plain of the Campi Flegrei, where activity has occurred on a much subdued scale ever since. Its name means "the burning lands", derived from the ancient Greek *phlegein* (to burn), although activity has been much reduced for many centuries.

Vesuvius from Naples, with the main cone of Vesuvius on the right and the ridge of Monte Somma on the left; between them lies the valley of the Atrio del Cavallo. The pale zone curving down slope from the Atrio marks the lava flow that invaded San Sebastiano during the eruption in 1944. The isolated white building at the left-hand base of the cone of Vesuvius marks the position of the old Hermitage and the old Vesuvian observatory. The Neapolitan industrial suburbs stretch out to the base of the cone, and the maritime passenger terminal of Naples occupies the foreground.

Vesuvius seen from the air from the southwest. The smooth paler cone of Vesuvius, with the western part of its crater in shadow, rises within the Somma caldera, whose northern rim forms the enclosing ridge of Monte Somma. The Atrio del Cavallo forms the semi-circular depression between the cone and the ridge, and from it issues the pale solidified lava flow erupted in 1944. Note the circle of villages at intervals all around the base of the Somma volcano and the more recent settlements that have encroached increasingly upon the volcano. In the southwest lie the larger coastal settlements, including Torre del Greco on the right, Ercolano (Resina) on the left and, farther on the left, Portici, which would probably be most vulnerable during the next eruption. (Photograph courtesy of David Wallace)

Meanwhile, the eastern cone continued to flourish and it formed what Earth scientists have called Somma volcano, after its most prominent relic, Monte Somma. However, during the past 25000 years, violent explosions destroyed the crest of Somma volcano and left behind a large hollow, or caldera, on its summit. The most recent of these great outbursts devastated, buried and immortalized Pompeii and Herculaneum in AD 79. The irony of the situation is that the Romans called this volcano Vesuvius, but it was still, in fact, Somma volcano, and the cone that we know as Vesuvius had yet to be born. It was between 787 and 1139 that eruptions began to build the cone of Vesuvius within the summit caldera of Somma volcano. Thus, the mountain is really two volcanoes in one: the older, and much larger, mass of Somma forms its broad plinth and is crowned by a huge saucer-like hollow in which the younger, smaller cone of Vesuvius lies like an inverted teacup. It would be more reasonable to call the dual volcano Somma–Vesuvius.

After 1139, Vesuvius rested for 492 years. The Campi Flegrei briefly took up the baton when Monte Nuovo erupted for a week in 1538. This little Renaissance eruption stimulated the first timid flowering of curiosity and enquiry about volcanoes in Campania. But, with the Renaissance had come the Protestant Reformation, and then the Catholic Counter Reformation that almost knocked scientific enquiry back into the Middle Ages. It was within this intellectual context that, just before Christmas in 1631, the volcano suddenly awoke from its slumbers with its most powerful outburst for over a millennium. The Spanish–Neapolitan establishment seized the chance to make Vesuvius a weapon in the Counter Reformation: the people should repent for the despicable sins that had so provoked the wrath of God.

The eruption in 1631 had a more lasting and crucial role in determining the behaviour of Vesuvius. It cleared the throat of the volcano, and, for the next three centuries, it was seldom dormant for more than a few years at a time. Thereafter, this almost constant volcanic activity, the extension of scientific enquiry beginning during the Age of Enlightenment, the revelation of the treasures of Pompeii and Herculaneum, and the popularity of the Grand Tour – all stimulated increasingly flourishing interrelationships between the people and their volcano. One of the many consequences was that the closer observation of Vesuvius helped lay down some of the important foundation stones of volcanology.

Scientific research showed that Vesuvius behaves in two main ways. Violent eruptions usually occur only after several centuries of repose; it is just as well that they happen so rarely, for they are lethal and highly destructive, as the eruption of AD 79 so amply demonstrated. The other type of activity has been in a much lower key. It characterized the behaviour of Vesuvius from about 787 until 1139, and from 1631 until 1944, when frequent eruptions of lava flows, ash and cinders in both periods were interrupted by dormant interludes lasting only a few decades at most. These eruptions varied in vigour, but they were all usually just powerful and beautiful enough to provide an awesome spectacle from Naples.

A colossal bibliography on Vesuvius, stretching back 2000 years, represents

Fumes rising from the eastern wall of the present crater of Vesuvius.

the most comprehensive biography of any volcano in the world, and one that only its Sicilian counterpart, Etna, could rival. Campania has been densely populated for 3000 years, and literate witnesses have observed the volcano ever since the Greeks first colonized the region. The recorded story starts when, in his famous letters, Pliny the Younger described the eruption of AD 79 in the best account of any eruption in the world written before the eighteenth century. During the past two centuries, Vesuvius has become the most scrutinized of volcanoes, and some of the most famous names in the Earth sciences have revealed and analyzed its behaviour. These investigations show no signs of abating: for example, the *Journal of Volcanology and Geothermal Research* devoted special numbers to Vesuvius in 1992 and 1998 and to the Campi Flegrei in 1984 and 1991; and sheaves of other academic studies have also been published besides.

The observers of these eruptions recounted different stories, told from different angles according to the intellectual and religious climate of the time, and to the extent of the scientific and technical knowledge available, and also according to their literary ability, for they had to rely almost exclusively upon word-pictures to convey their message. Sometimes, as in 1906, the scientific interest was paramount; sometimes, as in AD 79, it was the historical and archaeological importance; sometimes, as in 1631, it was the religious point of view that held sway; and at other times, notably during the eighteenth and nineteenth centuries, the eruptions provided fascinating spectacles for artists and authors, or even just for tourists, as ash poured down over arable land and vineyards, or lava flows, like pythons, seemed to swallow villages almost whole.

Since 1944, the volcano has undergone in its longest dormant period for over 300 years. The longer it remains dormant, the longer the molten rock beneath it can develop into an explosive cocktail, and the more destructive, dangerous and lethal the re-awakening will be. No-one can tell exactly when Vesuvius will spring back to life, but the experts have drawn up contingency plans to deal with the crisis that will inevitably ensue. The peril will be so serious that the plan is to evacuate all the people living in the most vulnerable areas before the eruption actually begins. Otherwise it will probably be too late to save them. Vesuvius is still the most dangerous volcano in Europe. It is only masking its hand.

Further reading

Guest et al. 2003; Kilburn & McGuire 2001; Nazarro 1997; Palumbo 2003; Scarth & Tanguy 2001; Sigurdsson 1999.

Chapter 2

Campanian volcanoes: in the beginning

Volcanoes dominated the story of Campania long before mankind first settled there, and Vesuvius now dominates the volcanoes. Mankind and the volcanoes together have created one of Europe's most striking landscapes.

Like all of the volcanoes of Italy, Vesuvius exists because Africa is colliding with Europe. Italy has borne the brunt of the collision between the African and European plates for many millions of years, and it has become a zone of both complex movements of the Earth's crust and of volcanic eruptions. When the plates in the crust collide, their edges sometimes break into smaller fragments that move about independently rather like ice floes jostle each other in a frozen river estuary. This is what has happened in the central Mediterranean area, where three main forces are acting in conflict. Along the southern edge of the European plate, the plate carrying the floor of the Tyrrhenian Sea is moving to the southeast, the small Adriatic and Apulian plates are moving westwards, and the margin of the African plate is pushing northwestwards and plunging beneath Sicily and Calabria in a vast subduction zone. The Tyrrhenian Sea is opening and widening, while the Adriatic Sea is being compressed. Caught between these pincers, the rocks of Italy have been folded, twisted, crumpled and fractured, and the peninsula as a whole has turned in a broadly anticlockwise direction. The movements of the plates are very slow (a few centimetres a year), but, as they have been going on for millions of years, they add up to quite extensive changes that have often caused major disasters. Southern Italy, in particular, suffers from very severe earthquakes and has witnessed hundreds of volcanic eruptions. Some of the most devastating earthquakes in Europe during historical time have shattered Sicily and Calabria: the latest such earthquake destroyed Messina and Reggio di Calabria in 1908; and, if the historical trends in that area continue, another disaster on a similar scale may not be far off. Among the volcanoes, Stromboli (in the Aeolian Islands) has been erupting for much of the past 2000 years,

Southern Italy: an image by courtesy of Dundee University Satellite station.

and almost continually at least since the end of the eighteenth century; Etna (in Sicily) is the largest and one of the most active volcanoes in Europe; and Vesuvius ranks with the most dangerous and violent volcanoes in the world.

Campania

Volcanoes have been erupting in Italy for well over a million years. They now extend in a line from Monte Amiata in Tuscany, through the lakes of Latium and the Alban Hills near Rome, to Campania, Monte Vulture, the Aeolian Islands, and to Etna in Sicily. The volcanoes north of Rome are probably extinct, whereas those to the south are visibly active. In fact, the

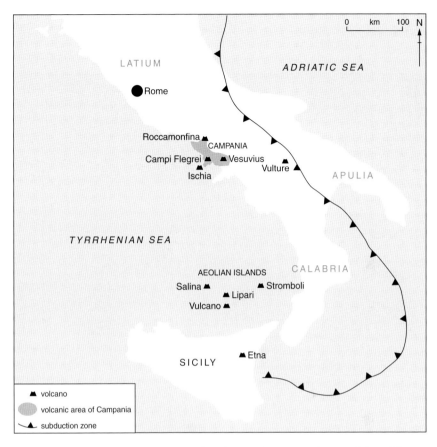

The major volcanic zones of southern Italy. The Earth's crust to the east of the subduction zone is plunging beneath the crust to the west of the subduction zone.

volcanoes of Campania were the youngest additions to this array, and they display the greatest variety of volcanic forms of any region in the peninsula. In terms of the geological timescale, Campania made its volcanic debut very recently – less than half a million years ago. On the other hand, in terms of the historical timescale, Campania has one of the longest periods of recorded history anywhere in the world, and volcanoes have played a dominant role in fashioning the human environment.

The eruptions in Campania were concentrated in three areas, although the volcano of Roccamonfina, in the north, has played only a minor role compared with the other two regions bordering the Bay of Naples. To the west of Naples lie the Campi Flegrei, whose name was given by the early Greek settlers. Soon, however, only fumes and bubbling mudpots remained to add

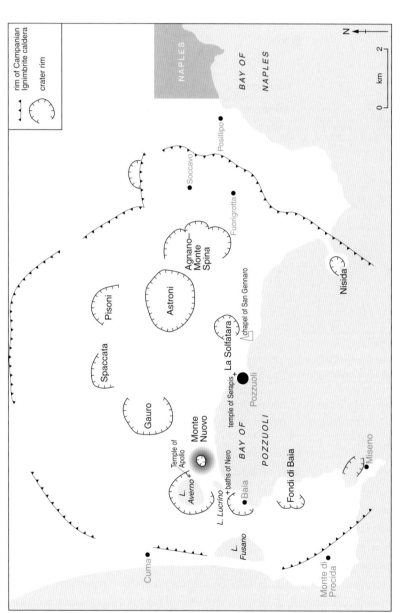

The volcanoes of the Campi Flegrei.

rim of Campanian
Ignimbrite caldera

crater rim

NAPLES

BAY OF
NAPLES

N

0   km   2

Posillipo

Soccavo

Fuorigrotta

Agnano–
Monte
Spina

Astroni

Pisoni

chapel of San Gennaro

La Solfatara

Nisida

Spaccata

temple of Serapis

Gauro

Pozzuoli

Monte
Nuovo

BAY OF

Temple of
Apollo

POZZUOLI

L.
Averno

baths of Nero

Fondi di Baia

L. Lucrino

Baia

Cuma

L.
Fusano

Miseno

Monte di
Procida

A mild emission from a fumarole, with typical sulphurous encrustations around the holes. The photograph is about a metre across.

a sinister air to the landscape where the Greek colonists had settled. To the east of Naples rises the double volcano of Somma–Vesuvius. It is just possible that the ancient Greeks could have witnessed one of the last explosive eruptions from Somma–Vesuvius before the volcano sank into the 700 year-old torpor from which it emerged so catastrophically in AD 79.

The Campi Flegrei volcanic field

Both the Campi Flegrei and Somma–Vesuvius began their volcanic lives in a similar fashion, but their behaviour then diverged so much that they now look completely different. About 50 000 years ago, the northern shores of the Bay of Naples may have looked even more beautiful than at present. Prolonged eruptions of molten lava had built up two large conical volcanoes, which rose gently to a high central crater that was situated over the main erupting vents. They would form a pair of superb cones, standing side by side, and rising to a height of about 1800–1900 m above the northern shores of the Bay of Naples.

About 35 000 years ago, the western (Campi Flegrei) volcano unleashed the most powerful eruption in the Mediterranean region during the past

The chief types of volcanic eruption

Everyone knows that volcanoes erupt lava, but that is only part of the story. They give off a vast range of products ranging from steam and gas to molten lava, ash, pumice and boulders, all of which are ultimately derived from magma, the molten material that rises from beneath the Earth's crust. Sometimes these products hiss or ooze from the ground, but at other times they explode from the crater with a fearsome violence rarely equalled on Earth. Any attempt to put eruptions into convenient groups involves much simplification. The standard texts on volcanoes in the bibliography provide more detailed information on the terms used to describe them, but, in a very simplified non-technical form, most eruptions can be grouped under four main headings: mild, moderate, vigorous and violent. In addition, an open-ended scale has been devised to describe the power of eruptions in succinct form, in which the weakest have a volcanic explosivity index of less than 3, and most severe eruptions reach 6 or, more rarely, 7.

Mild eruptions are often termed hydrothermal. They occur when magma lies close to the surface and heats the groundwater to boiling point. The heated water rises under convection to the surface, and can then emerge in hot springs, geysers, and mudpots. When fumes also emerge, the vents are called fumaroles. If they give off sulphurous fumes, they are called by their Italian name, solfataras. These mild exhalations commonly occur when magma is rising slowly towards the surface, often thereby offering warnings that more serious eruptions might be on the way. These features also occur frequently when a volcano is resting ("dormant"), when the magma still lies close to the surface between more active episodes. They have been common features for centuries on Vesuvius and in the Campi Flegrei.

Moderate eruptions are often called effusive–explosive. They mainly give off molten lava when very hot magma surges up the volcanic vent, or chimney, and reaches the land surface. If the magma contains little gas, then lava flows emerge and travel down slope across the land surface. If the magma contains more gas, the emissions of lava are mixed with explosions, which are caused when the pressure on the magma is reduced as it nears the surface. The explosions shatter the lava into fragments that cool to form cinders and ash, which then accumulate as volcanic cones above the vent. These cinder cones and lava flows are the most common eruptions on land throughout the world. These eruptions are also called Strombolian from the typical activity on the summit of Stromboli in the Aeolian Islands. Many eruptions of Vesuvius between 1631 and 1944 were of this type.

Vigorous eruptions are often more clearly explosive. They take place when large quantities of gas within a body of magma explode much more powerfully and noisily than in moderate eruptions. They commonly happen when groundwater is heated by rising magma and suddenly converted to steam within the confined space of rock fissures. The resulting explosions form hydrovolcanic or hydromagmatic eruptions that shatter some of the old cool rocks near the vent and scatter their fragments across the land. Dust, ash and cinders are ejected along with large blocks of rock, but lava flows are rare. These explosions have also been called Vulcanian, after the typical recent eruptions of Vulcano in the Aeolian Islands.

16

They often featured during the more vigorous activity on Vesuvius between 1631 and 1944. They have a volcanic explosivity index of 3 or less.

Violent eruptions are much more explosive. They occur when viscous, rather cool, magma containing much gas approaches the surface. The gases then blast out with an enormous force frequently exceeding that of nuclear explosions. These are the most powerful of all eruptions, and they are called Plinian, after Pliny the Elder, the chief victim of the eruption of Vesuvius in AD 79 – the first such eruption that was ever described – and after his nephew, Pliny the Younger, whose account of events in this eruption was unsurpassed anywhere in the world for over 1500 years. The explosions shatter the magma into the finest dust and ash, which are blasted far from the vent. They form huge billowing columns that reach the stratosphere and sometimes distribute volcanic dust all over the globe, and lower world temperatures for a year or two. At least two eruptions in the Campi Flegrei and several eruptions of Somma volcano have been truly Plinian in character, and a few more have been almost as powerful and have been termed sub-Plinian. Plinian eruptions are particularly lethal when they give rise to pyroclastic flows – or scorching glowing clouds of ash and fumes that race down hill, close to the ground, at great speed and destroying everything in their path. These eruptions have a volcanic explosivity index of 5 or 6, and even, very exceptionally, 7. The eruption of Vesuvius in AD 79 had an explosivity index of 5–6.

million years. It probably expelled up to $500\,km^3$ of glowing dust, ash and pumice in an enormous cloud, which formed great ashflows of phonolite. The ashflows smothered an area of $30\,000\,km^2$ over southern Italy and reached thicknesses of $60\,m$ in many places. This cataclysm might have occurred within a single tempestuous week, although it is also possible that the eruptions took place during brief disastrous episodes spread over a few thousand years. The ashflows then solidified and often welded together to form the Campanian Ignimbrite, which covers a large area of Campania. This great eruption wrecked the western mountain, and its summit collapsed by at least $700\,m$. A large central hollow – a caldera – formed in place of the summit, which was 12–15 km in diameter, and was one of the largest of its type in Europe. The remnants of the northern rim of the caldera can still be identified in an arc stretching from Monte di Pròcida in the west to Posillipo in the east. The edge of this caldera enclosed the area of subsequent volcanic activity that became known as the Campi Flegrei.

As time went on, the eruptions then focused more and more upon the centre of the Campi Flegrei. About 12 000 years ago, another violent outburst gave off almost $50\,km^3$ of ash, which formed the Neapolitan Yellow Tuff. Its fine fragments have become firmly welded together, and it now serves as one of the main building stones in the district. The eruption destroyed the central parts of the Campi Flegrei volcanic field and formed the Neapolitan Yellow Tuff caldera that sank by about 60 m within the Campanian Ignimbrite caldera. This second caldera is about 6 km in diameter; much of it lies below sea level.

Thereafter, volcanic activity in the Campi Flegrei became still more concentrated towards the centre of the Neapolitan Yellow Tuff caldera. The eruptions took place in three periods and became generally weaker each time. The first episode involved 34 eruptions between 12 000 and 9500 years ago, and their cones now form an arc stretching from Miseno in the west, around to the island of Nisida in the east. After a millennium of calm, six further eruptions occurred between 8600 and 8200 years ago, and they formed, for example, the deep craters of the Fondi di Baia north of Miseno. About 5000 years ago, the floor of the northern parts of the caldera was lifted about 40 m above sea level and formed the La Starza terrace, the distinct platform stretching northwestwards along the shore from Pozzuoli. This uplift of the land was probably related to an influx of magma into the crust, which did not rise high enough to erupt onto the land surface.

After the La Starza terrace had been uplifted, the third phase of activity took place between 4800 and 3800 years ago. These eruptions gave rise to some of the most prominent volcanoes in the Campi Flegrei today. Agnano–Monte Spina began the series 4800 years ago; Averno erupted about 4500 years ago; La Solfatara formed about 3800 years ago; and Astroni brought this

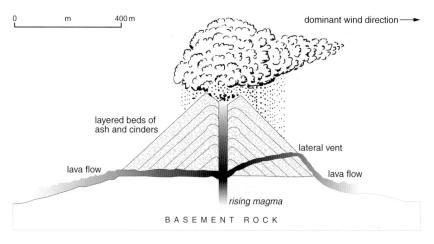

Aspects of a moderate effusive–explosive Strombolian type of eruption. Most of the airborne fragments accumulate and form a cone around the vent, whereas lava flows can extend for several kilometres from it.

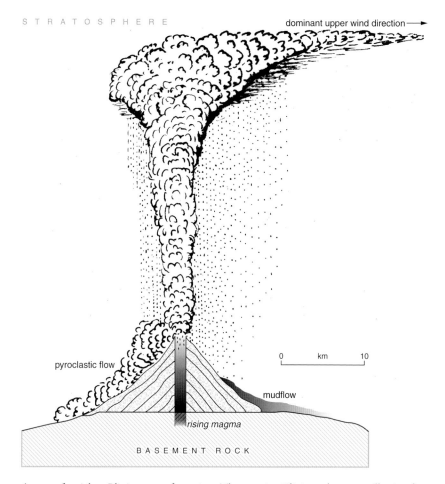

STRATOSPHERE

dominant upper wind direction ⟶

pyroclastic flow

0   km   10

mudflow

rising magma

BASEMENT ROCK

Aspects of a violent Plinian type of eruption. The towering Plinian column usually rises first; pyroclastic flows are often expelled as the eruption reaches its climax; the mudflows commonly develop at, or after, the end of the eruption. Airborne fragments are distributed very extensively. Lava flows are rare.

phase to an end. They often form wide craters, from 200 m to 1 km across, encircled by broad squat cones usually rising only about 100 m above their bases. They are typical of the cones formed during hydrovolcanic eruptions. Lava flows are very rarely formed.

The peoples of the Neolithic and Bronze Age probably witnessed some of these eruptions. They left no records of the events, although, as in many parts of the world, folk memories of their impact might have survived in their myths and legends. Such folk memories could well have been then recounted to the Greeks when they began to settle in the district in about 800 BC.

The rocks of Campania

Many very complex factors influence the nature of volcanic eruptions and the rocks that are formed from them. In Campania, some of the main factors that have influenced the types of rocks expelled and the style of volcanic eruptions have included: the chemical and physical nature of the magmas; the speed with which the magmas reached the surface or whether they rest for a while in reservoirs; and whether they encountered water on the way.

Specialist opinions vary about exactly how the magmas rise to the surface on Somma–Vesuvius. It seems that dark magma, broadly of basaltic type, rises continually from the depths into the crust. If the magma manages to reach the surface relatively unimpeded, it erupts during frequent episodes of persistent activity. These eruptions produce many lava flows, and also ash and cinders whenever the contained gases explode relatively mildly. These eruptions can build up a large mountain – such as the Somma volcano about 18 500 years ago.

However, if the magma halts in a reservoir, it undergoes both physical and chemical changes. For instance, it becomes cooler and more viscous; some minerals crystallize out from the mass and form layers of different composition, developing trachytes and phonolites, for example; and the gases contained within it tend to explode more violently when it approaches the surface. The longer the magma stays in the reservoir, the more differentiation can take place, and the more violent the subsequent eruptions can be. Such eruptions have expelled phonolitic materials from Somma–Vesuvius ever since the Mercato eruption. In AD 79, for instance, the magma erupted from a shallow reservoir, at least 5 km^3 in volume, which had been lying at a depth of between 3 km and 6 km for several centuries. At present, what could be a mass of magma has been identified at a depth of about 10 km.

In the Campi Flegrei, most of the original basaltic magmas seem to have paused as they rose towards the surface and have undergone differentiation in shallow reservoirs, so that the predominant volcanic rocks have been trachytes and phonolites.

Most of the magma mass beneath the Campi Flegrei has now cooled and solidified, but a small reservoir of molten trachyte, estimated at a mere 1.4 km^3, lies 4–5 km below the hub of the Campi Flegrei caldera at Pozzuoli. Unlike the lavas of Somma–Vesuvius, those of the Campi Flegrei contain no crystals of leucite, so it is relatively easy to distinguish between the lavas from the two volcanic areas.

In fact, the Campi Flegrei have lived up to their name for only a single week during the past 3800 years; thus, the name is not entirely appropriate. The best explanation is probably that many of the ancient volcanic sites have continued to exhale sinister and sometimes lethal fumes, and earthquakes have continually shaken the area ever since the ground rumbled beneath the feet of Virgil's Aeneas. The Greeks and their Roman successors believed that the Campi Flegrei marked the entrance to the Underworld. Odysseus and Aeneas both visited Hades from here, and at the Mare Morto ("dead sea"), near Misenum, Charon was reputed to row the souls of the dead across the Styx

Major eruptions in the Campi Flegrei

The violent explosions that gave rise to the Campanian Ignimbrite and the Neapolitan Yellow Tuff probably reached a volcanic explosivity index of 6. However, the eruptions in the Campi Flegrei during the past 10 000 years have been much less powerful than those of Somma–Vesuvius.

c. 33000 BC	Campanian Ignimbrite
c. 10000 BC	Neapolitan Yellow Tuff
c. 10000–7500 BC	34 eruptions in Campi Flegrei (e.g. Cape Miseno, Nisida)
c. 6600–6200 BC	6 eruptions in Campi Flegrei (e.g. Fondi di Baia)
c. 3000 BC	Uplift of La Starza Terrace near Pozzuoli
c. 2800–1800 BC	16 eruptions in Campi Flegrei (e.g. Agnano–Monte Spina 2800 BC; Averno 2500 BC; Monte Olbano 2500 BC; Astroni 1800 BC; La Solfatara 1800 BC)
c. AD 100–200	Bradyseismic movements around Pozzuoli
1198	Brief explosion at La Solfatara
1302 18 Jan.–March	Arso, Ischia
c. 1500–1538	Bradyseismic movements around Pozzuoli
1538 29 Sept.–6 Oct.	Monte Nuovo. Ash and cinders
1969–1972	Bradyseismic movements around Pozzuoli
1982–1984	Bradyseismic movements around Pozzuoli

to the Underworld. Hot springs made Baiae a leading spa; and the old crater forming the harbour at Misenum was a major base of the Roman fleet.

The single most evident testimony to volcanic activity in this period lies in the crater of La Solfatara ("the sulphurous") on the eastern outskirts of Pozzuoli. Eruptions began about 3800 years ago when the viscous lava extruded and formed the dome of Monte Olibano. Immediately afterwards, eruptions from a new vent shattered the northern half of the dome, and built the cone of ash and white pumice of La Solfatara alongside it. Some violent explosions then destroyed the southwestern sector of La Solfatara, and small pyroclastic flows surged forth and covered almost 1 km^2 towards the sea. The crater of La Solfatara has been the centre of hydrothermal activity ever since. Its sulphurous fumes reminded the ancients of the God of Fire, and thus the Romans called it the Forum Vulcani. Although La Solfatara never gave off lava flows, molten rock still lay close to the surface and heated the groundwater. But, despite the dark notoriety of the Campi Flegrei, the lakes, rich vegetation and varied relief made the area so beautiful that it became one of the most sought-after holiday resorts in the Roman world.

The growth of Somma–Vesuvius

Somma–Vesuvius is now the most violent volcanic child of the collision in Italy between the African and European plates. For thousands of years after

A pyroclastic flow descending the slopes of Mount Pelée in December 1902. The rapidly advancing leading edge of the cloud at ground level forms its chief destructive element.

great explosions had destroyed its companion in the Campi Flegrei, it formed a majestic cone rising from the northeastern shore of the Bay of Naples.

Somma volcano was built up by a long succession of mild emissions of lava flows that probably began below sea level more than 300 000 years ago. Its oldest violent tantrum so far discovered occurred about 25 000 years ago and has been called the Codola eruption. Somma soon resumed its more moderate ways and went on growing until it reached its maximum height about 18 500 years ago, when it formed a smooth cone, composed mainly of lava flows, rising to an apex about 1800–1900 m above sea level. It covered an area of about 480 km^2, with a volume approaching 150 km^3 – although it was still less than half the size of Etna in Sicily.

Pyroclastic flows

Pyroclastic flows are the most terrifying and lethal features in the volcanic repertoire and they commonly give rise to the greatest volcanic catastrophes. "Pyroclastic flow" is a term used to describe huge and sometimes glowing clouds of
scorching hot gas and volcanic fragments, ranging in size from dust, ash and pumice,
to large rocks, which are expelled at great speed in a turbulent aerosol-like mass.
Pyroclastic flows usually consist of two parts: a dense, ground-hugging base, the
pyroclastic flow proper; and a less dense, upper cloud, now called a pyroclastic surge.

The pyroclastic flow rushes across the ground at speeds commonly exceeding
150 km an hour, and sometimes at as much as 500 km an hour. Their temperature
varies between 200°C and 450°C. The less dense pyroclastic surge forms soaring
turbulent clouds of ash, dust, toxic gas and steam. It is these beautiful and terrifying
clouds that are seen in photographs, although the advancing prong of the dense
base forming the pyroclastic flow can sometimes also be detected. The ground-
hugging aspect of pyroclastic flows is their most lethal quality. During their unstoppable advance, they devastate everything in their path and even pick up fragments
of masonry, trees and vehicles, which they add to their armoury. Human beings
and animals are asphyxiated, burned, baked and bombarded by flying debris of all
sizes. Few buildings, and even fewer human beings, have ever been known to
survive their onslaught.

Pyroclastic flows come in several different forms. The most powerful of all are
those associated with a violent blast that is directed from the crater in a relatively
narrow sector. They can move at speeds of up to 500 km an hour and may cause
massive destruction. Such was the flow that destroyed Saint-Pierre in Martinique
in 1902. Less powerful are the flows generated when a huge erupting column loses
its upward momentum for a short while, collapses under the influence of gravity,
and surges down the slopes of the volcano at speeds often ranging between 250 km
and 400 km an hour. They cause very severe damage to buildings and can spread
more widely than those directed by blasts. The flows that destroyed Pompeii and
Herculaneum in AD 79 were of this type.

The least powerful flows often gush from a crater that has suddenly been enlarged by a major collapse of the summit of the volcano. They then travel down slope
under the influence of gravity at speeds of 150–200 km an hour. They too can cause
great damage to buildings, although strong major structures can survive their onslaught. Pyroclastic flows of this type seem to have erupted from Vesuvius in 1631.

Soon afterwards Somma changed its eruptive style. The long sequence of
mild effusions of lava came to an end, and they were replaced by a series of
brief violent eruptions, probably lasting no more than a week at a time, and
separated by long dormant periods lasting several thousand years. The first
eruption gave off such huge volumes of ash and pumice that the summit collapsed and formed a caldera, which was then widened and deepened by each
subsequent outburst. In total, these great explosions removed about 40 km^3

of material from Somma volcano after it reached its greatest height about 18500 years ago.

Modern Vesuvius later grew up within this caldera, and thereafter Somma was reduced to the role of providing a basal plinth for its smaller and now altogether livelier offspring, Vesuvius. The northern rim of the caldera stands out today as the ridge of Monte Somma, which curves in a protecting arm around the younger cone of Vesuvius. The eruptions from Vesuvius have now almost completely buried the lower southern rim of the caldera of Somma, although it could still be detected as a shoulder on the flanks of the mountain at the beginning of the twentieth century.

The eruptions of Somma–Vesuvius before 1631

The major prehistoric eruptions of Somma–Vesuvius, and that in AD 79, were violent and brief. They are known as Plinian eruptions in honour of the famous victim and his nephew who first described them. They are among the most powerful and destructive outbursts in the volcanic repertoire and commonly have a volcanic explosivity index of 5. Similar, but less powerful, outbursts, like the eruptions of 1631 and *c.* AD 472, have been called sub-Plinian and have a volcanic explosivity index of 4.

As far as may be discerned from archaeomagnetic dating and the brief references in the surviving texts, explosive outbursts of ash and cinders characterized the eruptions that took place between AD 79 and about AD 750. However, between AD 787 and 1139, fairly persistent activity occurred and emissions of lava flows seem to have become increasingly predominant. The dominant winds commonly distributed the ash and dust in a broadly north-easterly direction, and much less frequently towards the southeast. Major accumulations of airborne fragments in any other direction have been very rare.

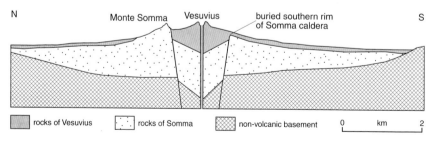

A simplified cross section of Somma–Vesuvius.

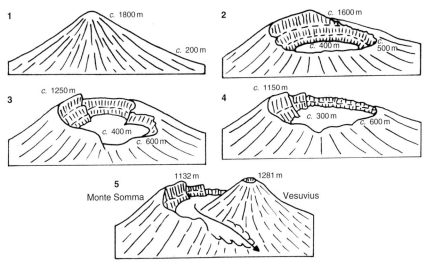

Stages in the growth of Somma–Vesuvius volcano

1. *Somma volcano in c. 16500 BC.* **2.** *The caldera formed after the Sarno eruption in c. 16300 BC.* **3.** *The caldera deepened and widened after the Avellino eruption in c. 1780 BC.* **4.** *The caldera after the eruption in AD 79.* **5.** *Vesuvius and Monte Somma at present, with the southern rim of the Somma caldera now hidden beneath lava flows and ash erupted from Vesuvius; the arrow shows the direction of many lava flows.*

The Sarno eruption

About 18300 years ago, a very violent eruption gave off masses of white pumice, followed by black cinders, which were thrown into the stratosphere and carried due eastwards by the dominant winds. Even as much as 45 km east of the volcano, the fragments accumulated to a depth of at least 1 m. This Sarno eruption was probably the most powerful in the whole history of Somma volcano; and at least 4.4 km³ of fragments were expelled. The eruption wrecked the crest of Somma, and such a volume of fragments was ejected that the western part of its summit collapsed, and formed a caldera 2 km across. Each of these momentous events probably took place in the space of a week or even less. The eruptions of Somma resumed their calmer course for some 2000 years when another vigorous episode threw out layers of the Novelle pumice about 16000 years ago. Another period of relative calm ensued for the next 8000 years, during which milder eruptions might have taken place within the caldera; but, if they did, no evidence of them can now be traced – because Somma blew up again.

25

A selection of volcanic fragments commonly expelled during violent eruptions. They range from a large smooth volcanic bomb (lens cap on top), to smaller rough clumps of cinders and to a carpet of much finer paler fragments of ash and pumice.

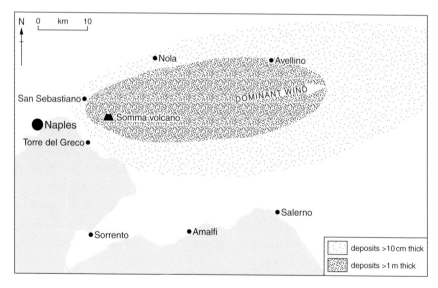

The distribution of airborne fragments erupted during the Sarno eruption.

The major eruptions of Somma–Vesuvius, 23000–700 BC

c. 23000 BC	Codola eruption (P?) (VEI 5?)
c. 16300 BC	Sarno (or Basal Pumice) eruption (P) (VEI 5)
c. 14000 BC	Novelle (Greenish Pumice, Pomice Verdolene) eruption (P) (VEI 5)
c. 6000 BC	Mercato (or Ottaviano) eruption (P) (VEI 5)
?	At least two undated eruptions (SP)
c. 1780 BC	Avellino eruption (P) (VEI 5)
?	At least three undated explosive eruptions (SP?) (VEI 4?)
c. 1000 BC	(S) with (SP) (VEI 4?)
c. 700 BC	(S) with (SP) (VEI 4?)

VEI = volcanic explosivity index (devised to compare the power of different eruptions)   S = persistent moderate effusive–explosive activity of Strombolian type   SP = sub-Plinian eruption   P = Plinian eruption
N.B. The types of eruption are not as clearly distinct as this table implies.

The eruption of the Mercato pumice

Volcanoes that start erupting violently often get into the habit. Somma suffered another violent eruption of ash and pumice about 8000 years ago, which Earth scientists have called the Mercato (or Ottaviano) eruption, named after the pumice found in the marketplace in Naples. The volcano first erupted white ashy pumice and then black cinders and pumice, and finally expelled pyroclastic flows as the eruptive column lost its upward impetus. Again, the fragments were thrown into the stratosphere, where the dominant winds carried them to the east-northeast. However, the Mercato eruption was less violent and gave out less material than its predecessor. Only between about 2.25 km^3 and 3.25 km^3 of ash and pumice were expelled, although they still formed a thick blanket, nearly 1 m deep, some 35 km east of the volcano. The eruption deepened and widened the caldera.

Further reading

Andronico et al. 1998, 2002; Baratta 1897; Cortini & Scandone 1982; De Vivo et al. 1992; De Vivo & Rolandi 2001; Giacomelli & Scandone 1992; Scandone et al. 1991; Scarth & Tanguy 2001; Spera et al. 1998.

Chapter 3

The Avellino eruption:
a prelude to Pompeii

Excavations have revealed relics of a small Bronze Age village, two human skeletons and the footprints of those who were trying to escape from this violent eruption.

The Avellino eruption has recently become the star disaster caused by the prehistoric Somma volcano. It was the early Bronze Age, and farming villages were already scattered quite densely around the flanks of Somma. This time, therefore, human beings would certainly witness the eruption. It also killed some of them, and two of its victims have been found; they were the first casualties ever to be discovered of a prehistoric eruption of Somma volcano. Archaeologists have since unearthed a village that has revealed invaluable details about rural life in the early Bronze Age. The Avellino eruption thus bequeathed to history a prelude to the great catastrophe of AD 79.

Two skeletons from the early Bronze Age

In 1970, at San Paolo Belsito, about 2 km from Nola, and about 18 km from the crest of the volcano, the skeletons of a man and a young woman were discovered buried beneath 70 cm of pumice that had been expelled during the Avellino eruption. As the pair fled from the volcano, the suffocating pumice had begun to rain ever more thickly down upon them. They had fallen to the ground and had died trying to protect their faces with their hands. The man was about 40–50 years of age, 1.7 m tall and quite robust, although he had suffered from rickets and arthritis in his knee. The girl could have been encouraging the man along as they fled. She was about 21 years of age and had been pregnant several times.

The Avellino eruption, about 1780 BC

This violent outburst of Somma–Vesuvius was named after the town of Avellino, 35 km northeast of the volcano, where some of its chief deposits are well exposed. Its erupted fragments cover pottery relics from the early Bronze Age in about 1780 BC. This violent eruption followed the now-familiar pattern. It began with explosions of fine ash, as the volcano cleared its throat, before the power of the eruption rapidly increased. Ash and first white, and then coarse grey, pumice were expelled in a column that towered 36 km into the stratosphere. The dominant wind winnowed the fragments out from the column and showered ash northeastwards down onto the sites of Nola, Benevento and Avellino. The crest of the mountain collapsed, widening and deepening the summit caldera even more, especially on its southwestern rim. At the same time, pyroclastic flows gushed from the vent, sped down the flanks of the volcano, and invaded the plains around Naples and Nola. About 10 km or 15 km from the vent, the pyroclastic flows began to cool and slow down, and the vast amounts of steam within them condensed. The waters mixed with the heavy rains unleashed by the storms that had been caused by the rising eruptive column. They combined to generate floods and mudflows that swamped the plains in small fragments 10–15 m deep that stretched as far as 25 km to the north and west of the vent. In fact, these pyroclastic flows seem to have played a more prominent role in the Avellino eruption than in any of its predecessors and they were probably the main cause of the devastation that the eruption produced. In all, about 4 km³ of fragments were expelled. The final stages of the Avellino eruption built a large cone of fine ash that overlapped the western edge of the caldera, about 3 km from the present cone of Vesuvius. Much of it has been destroyed by subsequent violent eruptions, although a remnant now forms the hill on which the Vesuvius observatory was built.

It seemed as if the couple must have met their deaths as they were fleeing from their village. It remained to find out where this village had been; this task was like looking for a needle in a haystack. Clearly, finding the village without digging up a vast area of concentrated settlements and good agricultural land would need a large stroke of luck, which came in May 2001.

An early Bronze Age village

In May 2001, the usual exploratory surveys were proceeding before a supermarket complex was built at Croce del Papa in Nola. The surveys brought to light an ancient furnace. Further excavations revealed that a mixture of volcanic ash and mud, 6 m deep, had moulded and preserved an early Bronze Age village, with a culture similar to that found in nearby Palma Campania.

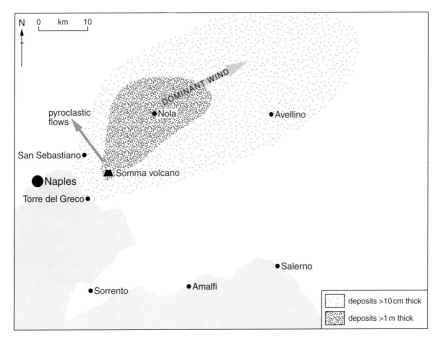

The distribution of airborne fragments and pyroclastic flows from the Avellino eruption in c. 1780 BC.

The little settlement was a far cry from the sumptuous palaces of Bronze Age Crete, but what it lacked in grandeur it made up for in the almost unprecedented details of village life that were revealed within it. The village might have had a population of about a hundred, and must have been one of many settlements scattered on the plain fringing the northern flanks of the mountain at that time. The villagers lived in horseshoe-shape huts made of wooden poles, tied together at their pointed crests, so that they looked rather like elongated Native American tepees. The three huts excavated were covered in thatch and straw, held together with mud and old volcanic ash. These huts were large enough to accommodate an extended family, including one room that measured almost 16×9 m. Like many homes in the Mediterranean area during the Bronze Age, each hut had a hearth and a storage area.

The women of the village were keen potters, weavers and basketmakers. The villagers owned dozens of ceramic utensils, most of which have been perfectly preserved. They range from large jars for storing nuts, wheat or barley, to small pots and eating bowls complete with their stands. The women also wove vegetable fibres into cloth. The men were farmers who grew subsistence crops of wheat and barley, and they had made a grain-threshing area

The remains of a hut, 9 m across, in the Bronze Age village near Nola preserved in the ash erupted during the Avellino eruption (courtesy of M. F. Sheridan and the Proceedings of the National Academy of Sciences, USA*).*

for the village. They used cattle both to plough the fields and to provide milk. However, the men were also hunter–gatherers: they picked nuts, and killed wild boars with arrows tipped with sharpened flint and bone. They also reared domesticated animals, such as sheep, pigs and goats, which they kept in pens enclosed by wooden and wattle fences. The remains of only a dozen or so animals, and no human skeletons, have been found in the village, which shows that the villagers must have fled when they saw the danger coming.

It is very unlikely that any of the villagers had ever seen a volcanic eruption, but they were certainly terrified. They were warned: the rumbling ground, the roaring explosions and the fearsome column billowing from the summit 18 km away would justify their panic-stricken exodus. Fine ash began to shower down on the whole area. A soothsayer might have had just enough time to accuse them of angering a god and bringing the catastrophe upon themselves. The villagers would be in no mood to linger: the danger was far too evident and far too pressing. In fact, the eruption seems suddenly to have taken a turn for the worse. It started to rain heavily, and the muddy ground preserved the footprints of the fugitives as they ran away, as well as those of their goats, pigs, sheep and cows. There was no time for leisurely appraisals of the danger, and one woman, for example, instantly abandoned the food that she had been baking in a jug in her oven. Nevertheless, the animals were valuable and had to be saved: someone opened the animal pens, released the beasts, and led them off across the plain, away from the hail of ash. However, in the general panic the villagers inexplicably abandoned thirteen goats in their pen; nine of them were pregnant, which would have made their loss even greater.

Many villagers must have had time to gather up their most treasured possessions, although the couple found at San Paolo Belsito had run off empty handed. However, someone left behind a head-dress made of sliced boar tusks – was it an oversight, or could it possibly have been a votive offering? The men would probably have taken their hunting gear, but there could be no question of anyone taking bowls of food or jars of grain. One of the dogs opted to stay put and cowered in one of the huts until the roof caved in under the weight of the ash and pumice and put it out of its misery.

The two victims found at San Paolo Belsito had struggled 2 km from the village, and might have been stragglers in the headlong race for salvation. The other villagers and their animals probably died a little farther from home. In any case, they would all have had to run more than 15 km to the north to reach a safe area where no fragments were falling. In the space of perhaps less than 24 hours, over 1 m of ash and pumice fell around Nola itself.

If the fugitives managed to survive the falling ash and pumice, the last

The skeleton of the young woman found buried beneath 1 m of pumice at San Paolo Belsito (courtesy of M. F. Sheridan and the Proceedings of the National Academy of Sciences, *USA).*

hours of the eruption reserved an even greater peril for them. The erupting column collapsed from time to time and the summit caldera widened and deepened as the crest of the volcano foundered. Pyroclastic flows surged down the northern and western flanks of the mountain, but by the time they had travelled 15 km or so from the vent, they had cooled and slowed down, and their vast steam content had condensed to water. Rain, too, fell in torrents from the storms generated by the erupting column. The soaking flows and the storms together drenched the ash and pumice clothing the slopes of the volcano and formed masses of wet mud that careered down the flanks of the mountain and rushed out in a great flood across the plain. The mud swept into the village, entombed all the goats, invaded every hut, sealed every nook and cranny, and buried every utensil, weapon and artefact. The mantle of mud preserved them all, as if in casts, for well over 3000 years as prime exhibits in what has been claimed to be the best-preserved Bronze Age village in the world. Perhaps some of the villagers who had lingered might possibly have managed to outrun the mud and the waning flows. Most of the villagers had already fled, leaving behind their footprints in the stiff ash.

Few survivors could have returned to their former homes: the villages and the fields around them were devastated for over 200 years.

Calm after the Avellino eruption

Relative calm returned to the volcano after the turbulent days of the Avellino eruption, and mild activity once more became the keynote of the behaviour of Somma. However, at least two explosive eruptions interrupted the monotony: the first occurred about 3000 years ago when grey cindery ash rained down to the east of the mountain; the second took place about 2700 years ago when ash and pumice were expelled to the northeast. Greek colonists would no doubt have witnessed this eruption – as well as any mild activity that might have continued for a while thereafter.

It seems that Somma soon returned to what had become its almost habitual period of repose after its exertions, for no Greek records of activity on the volcano have survived, if they ever existed. Nevertheless, some have suggested that the volcano may have erupted in 217 BC, although the evidence is flimsy. No deposits from this event have been identified, although one quite literally far-fetched tale about this supposed eruption claimed that it had expelled "stones" that fell on Rome. No other eruption from the volcano has

The footprints of two fugitives preserved in the ash expelled from the volcano during the Avellino eruption (courtesy of M. F. Sheridan and the Proceedings of the National Academy of Sciences, USA).

ever managed that. On balance, then, it seems that the volcano had been dormant for many centuries – perhaps almost a millennium – when it next awoke from its slumbers in AD 79.

Somma–Vesuvius just before AD 79

After a caldera has been formed, less vigorous eruptions commonly resume and expel lava flows, ash and cinders within it. Somma volcano followed the same pattern when a cone of ash and cinders grew up within the caldera. However, the Somma caldera was formed and modified by several very powerful eruptions, each of which blasted away the evidence of previous eruptions inside it. Thus, the present Vesuvius probably had several predecessors that were blown to smithereens by each successive violent outburst.

Unfortunately, during Greek and Roman times, descriptions and paintings of the volcano were too imprecise and ambiguous to provide any firm evidence whether a cone had formed in the midst of the caldera after the Avellino eruption. No author described such a cone, and no artist painted one. However, this is not as conclusive as it might seem, since the artists and intellectuals of the ancient world were not interested in producing accurate pictures or descriptions of the natural landscape. Consequently, scholars have not had much success in trying to find out exactly what the mountain looked like in AD 79. For example, one picture from the House of the Centenary in Pompeii (now in the Archaeological Museum in Naples) shows Bacchus standing next to a steep and jagged mountain that some have claimed to be Vesuvius. If it represents a real mountain at all, it must be said that it is quite unlike any volcano, let alone Vesuvius – although it does bear a passing resemblance to the decidedly non-volcanic Matterhorn in the Alps. However, a fresco depicting the loves of Venus and Mars (now in the Archaeological Museum in Naples) perhaps offers a more convincing picture. The background shows a mountain with a large central crater or caldera, surrounded by an even-topped rim with a pronounced notch that allows a glimpse inside the caldera. There is no cone within this hollow. This view could well show the state of the volcano before the eruption in AD 79.

Written accounts of the revolt of Spartacus offer a different picture. As the producers of Hollywood spectaculars know, Spartacus, the Thracian, led the revolt of some slaves and gladiators from the gladiator school in Capua in 73 BC. The Roman historian Florus (*Epitome of Roman history* II:8) recounted how about 70 men took refuge on the summit of the mountain and they were soon joined by thousands of other rebellious slaves. At that time, the summit area resembled a vast shell, about 1500 m across, with steep rocks tangled with wild vines, and thick woods infested with wild boars. The only access to the summit from the north side of mountain was by a narrow valley, which the slaves could effectively block and defend, because any attackers would have to approach virtually in single file. The southern side of the summit dropped away in an inaccessible precipice, so that it seems to have been very much like the present Monte Somma. However, no cone resembling the present Vesuvius apparently existed at the foot of this precipice. Instead, there was a plain that was extensive enough to house the tents of some 3000 Roman soldiers, under Claudius Glaber, when they came in pursuit of the fugitives. Spartacus ordered his men to make "ladders and ropes from the wild vines that covered the summit [of Monte Somma], and from the gorges on the edge of the hollow; and he made them descend by the rope ladders and ropes to the plain below . . . then the gladiators, turning by [both] the eastern and western slopes of the mountain, closed the Romans in a pincer, and killed them". However, the gladiators did not enjoy their glory for long. In 71 BC, a more astute Roman commander, Marcus Licinius Crassus, laid hands on the rebels. Spartacus himself seems to have got the thumbs down in a gladiatorial contest, although others maintain that he was crucified with his men on the Appian Way.

The implications of the Spartacus episode are that the old caldera had been filled with volcanic debris, so that it formed a plain stretching at the foot of the steep caldera edge of Monte Somma, and that there was no cone resembling the present Vesuvius on this plain – otherwise the Roman army would not have had room to camp there. Therefore, the cone of Vesuvius must have been formed at the foot of the Somma scarp *after* the cataclysm in AD 79.

The balance of the indications would suggest that there was no central cone, and that neither a prototype of Vesuvius, nor the present Vesuvius featured in the caldera on the eve of the catastrophic eruption in AD 79. Thus, the central cone must have grown up only during the next millennium and would become known to modern observers as Vesuvius. The irony of the eruption of AD 79 is that the mountain that the Romans knew as Vesuvius was, in fact, the much older, decapitated, but still substantial Somma volcano. Our Vesuvius had yet to appear.

Further reading

Albore Livadie 1986, 2002; Mastrolorenzo et al. 2006; Florus 2.8; Plutarch *Crassus* 9.1–3.

Chapter 4

The eruption in AD 79: the day of wrath

The eruption in AD 79 is the most famous in the world, because it destroyed, buried and immortalized Pompeii, Herculaneum and other Roman settlements in Campania. Archaeologists and Earth scientists are still bringing its enormous legacy to light.

The eruption in August AD 79 was by far the grandest, most lethal and most destructive outburst of Somma–Vesuvius during the past 3000 years. Everyone knows that a thick shroud of ash and pumice buried Herculaneum, Pompeii and many smaller settlements around them, and bequeathed to them an immortality that has few parallels in history. They remained entombed for 1700 years and more, until treasure hunters and archaeologists slowly began to reveal their secrets to an amazed and admiring world. Indeed, the resurrected settlements have achieved far greater fame than they ever had when Roman citizens walked their streets. Thus, the eruption destroyed – and preserved – the most famous archaeological sites in the world, and they, in turn, have helped make Vesuvius the most famous of all volcanoes.

Roman Campania

In Roman times, Campania was a wealthy and densely populated region, where agriculture provided the main basis of a network of flourishing trading centres. Naples was the hub and chief city in the area, with a population of perhaps 50 000. Capua, in the north, Nola, to the northeast, and Pompeii, lying to the southeast, were three of the main market towns serving the agricultural districts around them. Each may have housed as many as 20 000 people. Stabiae, Herculaneum, Pozzuoli, Baiae, and Misenum, each with 5000–6000 inhabitants, formed a chain of thriving settlements along the shores of the Bay of Naples.

Campania was lauded by the Roman poets and cherished by the Roman aristocracy. Pliny the Elder had called it *Campania felix*, because it was one of the most beautiful regions in the empire. The Roman historian Florus voiced the general view when he claimed that Campania was "the fairest of all regions, not only in Italy, but in the whole world". The air always seemed light, pure and healthy: the Apennines protected Campania from the cold north winds in winter, and the northwesterly breeze from the Mediterranean Sea tempered the heat in summer. The dark-brown volcanic soil was rich in humus and renowned for its fertility, and the same plot of land would often yield two crops a year. The grass was always green, and farm animals flourished on the lush pastures. Wealthy and able owners brought the best out of their rural environment. The Romans had developed a fine network of aqueducts and channels for drainage and irrigation. War veterans had been allocated land, which they often cultivated with some expertise. The large country estates were also linked by a network of good roads that ensured rapid export of produce to the market towns. Luxuriant vineyards and market gardens spread almost to the very crest of Somma–Vesuvius.

These fertile soils and the rocks beneath them carried a warning that the people could not comprehend. Campania was almost entirely floored by volcanic rocks that had erupted within the past few thousand years. Pompeii itself stood on the snout of a lava flow from Somma–Vesuvius and mild emissions of steam and fumes still issued from innumerable holes and vents, especially in the area west of Naples. In antiquity, the steam and fumes in the Campi Flegrei led the Roman poets, whose imagination far surpassed their knowledge of active volcanoes, to believe that they marked the entrance to the Underworld. Nevertheless, Campania could bear no comparison with Etna in Sicily, or Stromboli in the Aeolian Islands, both of which often seemed to erupt molten rocks a-plenty. The volcano that they called Vesuvius had not erupted such molten rocks for many centuries. It is, indeed, rather ironic that Pliny the Elder did not include it on his list of known active volcanoes when he published his famous *Natural history*. The classical authors had no concept of active, dormant and extinct volcanoes, and Pliny the Elder might have simply omitted the mountain from his list because it had not erupted within living memory.

At that time, only a few scholars recognized the mountain for what it once had been. Diodorus Siculus, in his *Library of history* (4.21) written in the first century BC, said that Vesuvius displayed many vestiges of "ancient fires" that could be compared to those being expelled by Etna in his own day. Vitruvius, in his treatise *On architecture* (2.6), written at the beginning of the first century AD, declared that the volcano had previously expelled flames from its mouth and had swamped the surrounding fields with pumice and materials that had been reduced to a similar state by the fire; but he did not say whether he believed that Vesuvius might ever erupt again.

The Greek geographer, Strabo (64 BC to AD 25), who wrote his *Geography* (4.5.8) in about AD 7, had probably seen the volcano with his own eyes:

> Dominating the area is Vesuvius, which is covered with fine fields except on its crest. A considerable part of the summit is flat, but all of it is unfruitful, and looks ash coloured. Its rocks are the colour of soot, with holes in them like pores, so that they look as if they had been eaten by fire. Thus, it can be inferred that this district was on fire in former times and had craters of fire that were quenched only when the supply of fuel gave out.

Strabo's remarks suggest that most scholars at the time believed that "the fire" had ceased "to burn", and that, in modern terminology, Vesuvius was dormant or extinct.

The Roman Empire in AD 79

All things considered then, Campania was a place that made relaxation seem a virtue. Many Roman aristocrats had built their opulent villas all along the scintillating shores of the Bay of Naples, where the views often extended from Misenum, out in the far west, to the distinctive bulk of Vesuvius closing the eastern horizon. On hot summer days, the mountain disappeared into the shimmering blue haze that formed over the bay almost as soon as the Sun had risen behind it. Here, the richest and most powerful people on Earth relaxed and plotted their next coups. The Emperor Augustus had died in AD 14 on one of his family's properties at Somma–Vesuviana on the northern slopes of the volcano. Julius Caesar, Cicero and the Emperor Nero were among other famous people who had lived here. So, too, had Nero's mother, Agrippina, until her ungrateful offspring had her murdered in his Campanian palace.

In AD 79, the Roman empire was approaching its apogee, although it might not have been as obvious to its citizens at the time as it seems with hindsight today. Just over a decade previously, it had looked as if the empire was foundering into chaos. In AD 68, the Imperial guards had assassinated the Emperor Nero and brought his lunatic antics to a welcome end. Then, during AD 69, no fewer than four emperors struggled for the Imperial diadem. The surprising outcome was that Vespasian, the best candidate, prevailed. He is one of those emperors whose features are easily recognizable from his busts in museums, for he looked like a kindly chubby-faced shepherd. During a reign lasting just six days short of a decade, he proved his mettle and gave peace to his vast dominions, and peace of mind to his subjects.

The Emperor Vespasian died on 23 June AD 79. His eldest son, Titus, succeeded him. Titus was 39, and had already been sharing his father's official responsibilities for several years. Titus had been a military commander – he had destroyed the Temple in Jerusalem in AD 70 during the Jewish wars – but he had also shown considerable political and social acumen. On his accession, Titus was clearly as sensible and capable as his father; with the same benign expression, but, as yet, without the paternal chubbiness. His only peccadillo was that he had fallen in love with a local princess, Berenice, when he had been fighting the Jewish wars. Even then, his good sense had prevailed, and he had renounced her, and he refused to succumb to further temptation, even when she came to Rome. Titus seemed to be too good to be true: he was an emperor who wanted to help his subjects. As Dio Cassius later explained, "after Titus became ruler, he committed no act of murder or amatory passion, but showed himself to be upright and self-controlled . . . He put no senator to death, nor indeed was anyone else slain by him during his reign." He even went so far as to institute measures to make people's lives more secure and free from trouble, and he banished informers from Rome. He himself was frugal and he indulged in no unnecessary expenditure. No cynic would be surprised that such a paragon reigned for a mere two years, two months and twenty days before he succumbed to a suspicious death.

Titus visited Campania just after his accession. He would almost certainly have gone to Naples, and he probably also went to Herculaneum. There, his doctor Apollinaris achieved a form of immortality for himself by inscribing on the walls of a latrine in the House of the Gems that he (Apollinaris, not Titus, of course) had enjoyed a good bowel movement: "*Apollinaris medicus Titi imp. hic cacavit bene*". Suitably refreshed, the Imperial party would most probably have gone on to inspect the Roman fleet lying at anchor at Misenum. Emperor Titus then presumably made his way back to Rome to undertake yet more good works. His reign did not prove to be a happy one.

42

The Pliny family

The commander of the Roman Imperial fleet based at Misenum was Gaius Plinius Secundus, who was to become known to history as Pliny the Elder (AD 23–79). The Pliny family came from Como, in the Italian Lakes region at the southern foot of the Alps. They had wealth, brains and power in abundance, and Pliny the Elder was one of the leading intellectuals of the age. He had an insatiable curiosity about the natural world and, whenever he travelled, he always kept a secretary with him to note down his incessant observations. He slept little and combined high intelligence with untiring industry and great powers of concentration. He wrote books on Roman history, cavalry tactics, grammar and oratory. Most important of all, in AD 79, he had just completed his famous *Natural history* in 37 volumes, a work that was to form a basis of the natural sciences in Europe until the close of the Middle Ages. On the face of it, Pliny the Elder was an unlikely admiral of the fleet, although he had in fact been a financial procurator in Spain and had spent 12 years as an army commander in Germany, where he became a friend of the future emperors Vespasian and Titus. Thus, Pliny the Elder was not only very rich but also well connected. At 56 years of age, he was at the height of his considerable powers, although he was becoming rather overweight and showing signs of breathlessness. He was also about to become the most famous of all the thousands of victims of the world's volcanic eruptions.

Pliny the Elder's sister, Plinia, had come to spend the summer with him at his splendid villa overlooking Misenum. She had brought her 17-year-old son with her, Gaius Plinius Caecilius Secundus, who became known to history as Pliny the Younger. He was born in Como in AD 62, and was the second son of Lucius Caecilius Cilo. Both his father and his elder brother died when he was a child, and he was therefore mainly brought up by his widowed mother, Plinia. The commander of the fleet was unmarried and, in AD 79, he had just adopted his young nephew. He had set about organizing his education, so that the lad could soon take up a position in the Empire that befitted his status. Thus, when Vesuvius erupted on 24 August AD 79, the diligent student was hard at work studying Livy's *History of Rome*, and taking little notice of the Imperial fleet assembled in the harbour at Misenum, and quite oblivious to the glorious view of the Bay of Naples beyond it. He could never have guessed at that moment that he would compose for posterity two incomparable accounts of the greatest natural catastrophe that afflicted Italy during the era of Imperial Rome.

When he was 18, Pliny the Younger began his career at the bar and managed to survive the murderous whims of the Emperor Domitian. He then reached high office in the administration, after his friend Trajan became emperor in AD 98, and he became consul in AD 100. At the same time, he became an accomplished orator and essayist on contemporary affairs, ranging, for example, from his wife's miscarriage, to a eulogy to the poet Martial, as well as on the eruption of Vesuvius. He was well known as a wealthy benefactor to the poor and the founder of a library in Como. His first two wives died young, but he enjoyed a happy marriage with his third wife, Calpurnia. In AD 110, Trajan appointed him governor of the province of Bithynia in Asia Minor, where he died at the age of 51 in AD 113.

The two letters of Pliny the Younger

As evidence of environmental events – as opposed to works of literature – accounts of eruptions in antiquity are too brief and too fanciful to be at all reliable. Some of them are poetic, most are entangled with tales of gods and mythical beasts, and many are crazy. But Pliny the Younger's two letters about the eruption in AD 79 are of an entirely different calibre. His vivid descriptions shine out like beacons of clarity among all the accounts of eruptions that have come down from antiquity. They are masterpieces.

Some 25 years after the catastrophe, when the Roman historian Cornelius Tacitus was preparing his *Histories* of Rome, he asked Pliny the Younger to give him an accurate description of the events leading to the death of his renowned uncle during the eruption. Knowing that his account would form part of a sober and reputable work of history, Pliny the Younger wrote Tacitus a letter in which he evoked the glorious death of the great man. He described every incident that he had either seen himself or that reliable witnesses had told him immediately afterwards, when events had been still fresh in their minds. This letter whetted the historian's appetite, because Tacitus then asked Pliny the Younger to describe his own terrifying adventures during the eruption when he was at Misenum. Pliny the Younger thus wrote a second, less formal, letter, although he realized that Tacitus might not need it for his historical study.

Pliny the Younger composed both letters in about AD 105–106. Unfortunately, the books of the *Histories* in which Tacitus detailed these events are now lost. However, both of Pliny the Younger's accounts have survived as

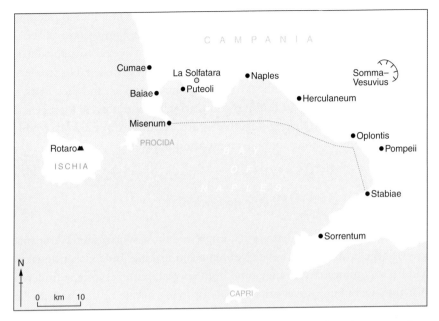

The Bay of Naples in Roman times, with the course of Pliny the Elder's last journey from Misenum to Stabiae.

Letters 16 and 20 in Book 6 of his published *Letters*. He did not set out to provide Tacitus with a complete picture of events. He makes no reference to the destruction of Pompeii and Herculaneum, nor to the fate of Naples, because he would realize that Tacitus would already have ample evidence about them. Pliny the Younger never went near Pompeii during the eruption, and in fact, it is unlikely that he had ever set foot in it. Pompeii was scarcely the kind of place that an aristocrat like Pliny would frequent.

The description of the eruption itself was not the prime aim of either letter, because such a task would have seemed superfluous and uninteresting in antiquity. Nevertheless, Pliny the Younger produced a vivid account of the progress of the eruption as it affected him and his two closest relatives. He was an accomplished letter writer, with a keen eye for detail and a thorough knowledge of literature. He also scrupulously revised and amended his words before they were published for posterity. It is by no means easy, even for a trained modern Earth scientist, to analyze the seemingly chaotic events of a large and terrifying eruption. But Pliny the Younger was a perceptive and highly intelligent young man. And his descriptions are all the more remarkable because he had no training in science, and had no methodological basis upon which to analyze what he was observing and fit it into a coherent pattern. It is a measure of his achievement that volcanologists and archaeologists have recently been able to match his account of events with the detail of the different layers of ash and pumice that Vesuvius expelled. His two letters gave volcanology one of its earliest and finest bonuses, and they mark a high point of volcanological observation, which was not even approached again until the end of the eighteenth century. All this from a lad of 17 who had never seen, let alone studied, an eruption in his life.

Damaging earthquakes

The only environmental hazard that impressed those who visited Campania was that the ground was forever wobbling and shaking. Indeed, earthquakes had been common there since time immemorial, but everyone considered that they were but a small price to pay for living in one of the most idyllic sites in the whole empire. Luckily, most of these earthquakes caused little damage to buildings or injuries to people. However, very occasionally the ground quaked with disastrous effects.

One of the most damaging of these Campanian earthquakes in Roman times took place on 5 February AD 62. Both Seneca (*Natural questions* 6. 1.1–2, 6.27.1, 6.28.1, AD 62–63) and Tacitus (*Annals* 15.22, published AD 120) make brief references to the event. The epicentre of the earthquake was probably located in the area lying to the southeast of Vesuvius, around Pompeii itself, where the damage was so substantial that scarcely a building was left intact. This was probably because alluvial silts alongside the River Sarno would magnify the subterranean vibrations and hence cause greater destruction, which, in fact, could have reached point IX on the 12-point modified Mercalli scale of earthquake intensity.

The moment of the earthquake was actually portrayed in stone in a remarkable bas-relief that was excavated in Pompeii in 1876 and is now in the Archaeological Museum in Naples. It was discovered in the household shrine in the home of Lucius Caecilius Jucundus (a banker) and it shows the Temple of Jupiter and the triumphal arch in the Forum in Pompeii.

Seneca wrote that:

Pompeii has been ruined by an earthquake, and the surrounding regions damaged too . . . The earthquake occurred on 5 February, and devastated

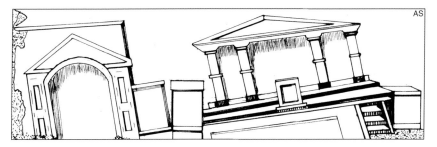

Sketch, based on a Pompeian bas-relief, showing the tilting of the buildings during the earth-quake on 5 February AD 62. The buildings were also carved asymmetrically to emphasize the effects of the motion.

with great destructiveness the whole of Campania, which has never been totally free from this curse, but has been undamaged and has often got over its fear. For part of the town of Herculaneum has collapsed, and even what remains stands precariously, and the colony of Nocera, though not ruined, still has things to complain about, and Naples has a lot of damage to private property, though not to public buildings, being let off lightly by the great disaster. Villas have been destroyed, and elsewhere have been shaken but not harmed. There are also these reports: a flock of sheep was killed, and statues were split in two; afterwards some people wandered about deranged, in a state of shock.

Suetonius (*Lives of the twelve caesars*: Nero) recorded a smaller earthquake in AD 64 that affected the Naples region. Suetonius claimed that Nero had been singing in the theatre in Naples at the time, and that the emperor, who had an unfailing belief in his own talents, had continued his performance unabashed and undeterred. However, that more reliable witness, Tacitus (*Annals* 34.1), asserted that the theatre fell down only after the Imperial performance had come to an end – and, in fact, he does not explicitly state that an earthquake had made it collapse.

It is not clear whether these earthquakes were caused when magma first began to rise beneath Somma–Vesuvius, because they were not apparently centred beneath the mountain and they took place an unusually long time before the eruption. Thus, they were more probably repercussions of the collision of the European and African plates. However, Pliny the Younger noted that increasingly frequent weak earthquakes had affected Campania in the days just before the eruption. These shocks would be more clearly linked to magma rising from the depths, which probably also had the effect of pushing up the land around the volcano. The shoreline at Herculaneum, for instance, was raised by 0.50–1 m just before the eruption, so that the waves could no longer reach and carry away rubbish that had been dumped on the beach.

The earthquake of AD 62 had shaken down many buildings in Pompeii, fewer in Oplontis and Herculaneum, and hardly any as far west as Naples. It is said that some of the more timid members of the population left the area to settle in what they hoped would be more stable zones (They would, in fact, have had to travel some distance, for such safe zones would be hard to find in earthquake-prone Italy). Some buildings were abandoned by their owners, including two of the most beautiful homes in the whole area: the Villa of Mysteries on the outskirts of Pompeii, and the Villa of Poppaea Sabina at Oplontis. Nevertheless, when the eruption buried the town 17 years later, most of the public buildings had been restored to their old eminence, although repairs to many private buildings were incomplete and their tottering walls were still shored up. Many houses had also been adapted to other functions. The wealthy Julia Felix, for instance, let off most of her vast property for conversion to boutiques. Some of the buildings were sold off to what would now be called property developers. Ruined houses of poor quality were knocked down altogether and planted with orchards, vineyards, and market gardens; part of the damaged House of the Fruit Orchard was taken over by a tradesman; bath houses were abandoned and turned into store rooms; and a freed slave called Sextus Pompeius Amarantus transformed a desirable residence into an inn. But the town had never lost its role as the bustling centre of its rich agricultural region.

Pompeii

Pompeii stood at a height of about 25–30 m above sea level, overlooking its small port at the mouth of the River Sarno. To the south, a low plain fringing the Bay of Naples stretched out as far as Stabiae at the foot of the rugged white limestone hills of the Sorrento Peninsula. To the north, fertile and gently sloping agricultural land rose well up the flanks of the volcano.

The walls of Pompeii enclosed an area of 1500 m by 1000 m, but they had been more or less useless since the Romans had pacified the region in 90–80 BC. However, the walls had retained a dozen defensive towers and seven imposing gates, which carried the main roads out into rural Campania.

Within the town, the streets followed the usual Roman grid pattern; and the main arteries were 8.5 m wide, with high pavements to keep pedestrian feet dry when it rained. The streets were commonly floored with large slabs hewn from quarries excavated into the ancient black lavas on the mountain slopes. They were an unheeded warning of the threat from that striking landmark lying just 10 km away on the northern horizon. On the other hand, the Pompeiians probably thought that the ruts worn by the carts and chariots into the slabs presented a far more urgent problem, especially since municipal elections were taking place. Indeed, many inscriptions and graffiti supporting the various candidates for these elections were decorating the streets during the summer of AD 79.

Pompeii was a busy place, with a population of about 20 000, of which some 8000 were slaves. The citizens lived in brick houses, one or two storeys high, with monotonous facades masking rich and varied interiors that were often built around colonnaded courtyards and gardens. The homes of the wealthier citizens were decorated with wall paintings of rural, mythological or erotic scenes, which commonly had a background of the famous "Pompeiian red". For instance, the House of the Vettii, owned by a rich freed slave, was covered in fine paintings. Other citizens decorated their homes with family portraits: one of the most famous of these is the picture of the

The wheel-rutted streets in Roman Pompeii, paved with slabs of lava quarried from flows previously erupted by Vesuvius.

baker, Terentius Neo, and his wife, who now stare back at their admirers in the Archaeological Museum in Naples. Many houses contained bronze or marble statues and mosaics; but none could compare with the fabulous mosaic in the House of the Faun showing Alexander the Great about to capture Darius, the King of Persia, at the Battle of Issos in November 333 BC.

Several temples catered for an array of tastes and deities, both Roman and foreign, and those dedicated to Jupiter, Apollo, Isis and Venus, for example, formed some of the most impressive buildings in the centre of the town. But the focus of Pompeii was the Forum, which chariots were forbidden to enter. It formed an open rectangle, 142 m long and 38 m wide, surrounded by temples and marble colonnades. Here, away from the constant rumble of the chariots, the more notable citizens could discuss the affairs of the day in relative peace, if not harmony. From time to time, they might have glanced up at the graceful mountain on the northern horizon; but they probably never gave it a second thought.

Although the town was quite compact, it still contained extensive areas devoted to gardens, vineyards, orchards and smallholdings, for growing cherries, peaches, almonds, onions, cabbages, chickpeas and walnuts, as well as for rearing pigs and poultry. Trade and business flourished practically everywhere. There were tanneries, banks, laundries, groceries, greengroceries, drapery shops, jewellers and fishmongers, not to mention 35 bakeries and flourmillers, a large market, a weights and measures office, and a tower from which water was distributed from the aqueduct to the public fountains in the streets. There were even dealers in slaves and property, who took a 1 to 4

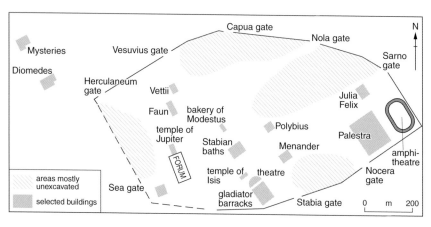

The town plan showing some of the main buildings revealed in the excavated areas of Roman Pompeii.

per cent commission on each transaction. Specialist craftsmen worked in the woollen textile workshops, combing, spinning and fulling the cloth, using urine that the citizens could contribute whenever they felt the need. Other specialists made large storage jars that Pompeii exported all over the country, chiefly from the port on the River Sarno. These jars often served to export the highly regarded local wine. Leisure activities were varied. Like most Roman towns of similar size, Pompeii had as many as 89 taverns, which often fulfilled the functions of bistros; a large gymnasium for those who felt the need for exercise, three large and well appointed public baths, plus schools, restaurants and stables. Most of the brothels were run in little rooms "above the shop", but one establishment was a more professional affair, decorated with helpfully explicit wallpaintings; here, prices ranged from two to four times the cost of a measure of wine. For the more passive citizens, the town had cockfight booths and two theatres. The larger, open theatre could seat 5000 people who came to see the best comedies of the day; the smaller covered building catered for up to 1300 people who preferred poetry recitals and musical performances. The nearby amphitheatre was much larger than either, and could seat 12000 spectators for its gladiatorial contests. Feelings during these contests seem to have ridden high. When the Pompeiian gladiators met those from neighbouring Nocera in the amphitheatre in AD 59, hooligan elements on both sides engaged in a bloody battle that resulted in deaths and injuries. The sober Tacitus disdainfully commented that these things were typical of such disorderly country towns. The Roman Senate closed the amphitheatre for ten years. However, in AD 62, either the earthquake or perhaps the intervention of Nero's new wife, Poppaea Sabina, induced the Senate to relent and allow gladiatorial contests to resume.

Herculaneum

Herculaneum was older than Pompeii, but much smaller. It had originally been a Greek settlement, and Hercules in person was reputed to have founded the town in the days when living legends roamed the world. In AD 79, Herculaneum was a prosperous little place with a population of about 5000, set on a low headland, 15–20 m above sea level, with fine views westwards to Naples and beyond. Somma–Vesuvius itself blocked the eastern horizon, some 7 km away. Herculaneum specialized in craft industries such as spinning and weaving, mosaic making and marble working, as well as fishing. The public buildings and private villas were generally finer than those in Pompeii, with delicate mosaics, fountains and statues. One of the largest buildings in the town housed a priceless collection of 1800 Egyptian rolled papyri written in Greek or Latin, which the latest techniques of multispectral digital imaging may yet save from oblivion; some experts hope that their charred remains might possibly include the texts of classical works that had been previously thought to have been completely lost. In another house was found a cross that may have belonged to a Christian converted by Saint Paul when he had preached in Pozzuoli in AD 61, although Herculaneum probably had far more adherents to the Epicurean philosophy of securing happiness whenever the chance presented itself.

Herculaneum was one of several similar settlements scattered along the shores of the Bay of Naples and over the lower flanks of the volcano. They included a vast villa at Oplontis, which had belonged to Poppaea Sabina, and the villa of Pomponianus, a friend of Pliny the Elder, at Stabiae. Most of the settlements lying within a radius of 15 km of the summit of the mountain had been built on volcanic rocks spewed from the crater in the distant past; another outburst would put them in grave danger. When the eruption came, a wide area, especially to the southeast of the mountain, was covered in thick ash and pumice, which is still probably enshrouding most of its victims and their homes. Only a small fraction of that area has been excavated, but luckily

Wooden beams charred by the pyroclastic flows at Herculaneum.

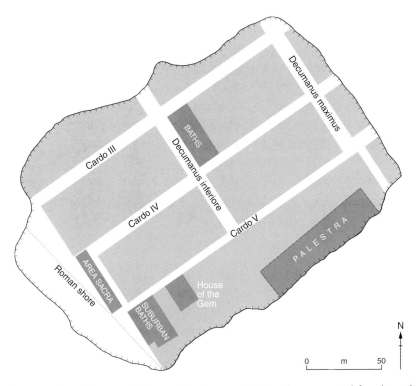

The town plan of the parts of Roman Herculaneum that have been excavated from beneath the modern town of Ercolano (Resina).

it includes three-quarters of Pompeii, perhaps a quarter of Herculaneum, and some large villas, including those at Oplontis, Boscotrecase and Stabiae.

The southern flanks of Vesuvius: 24 August AD 79, morning

Pliny the Younger did not describe the beginning of the eruption, because he did not see it. He was 32 km away at Misenum. The only surviving text recounting the early phases of the eruption is found in book LXVI, 21, of Roman history, by Dio Cassius. The work has made a tortuous journey into modern libraries. Dio Cassius was writing in Capua in AD 202–203 and published his work between AD 207 and AD 219. His account of the eruption of AD 79 is thus based on older texts that have not survived; neither in fact has Dio's original text, which has come down to modern times in a Greek summary that was made by Xiphilinus, a monk who lived in Constantinople in the eleventh century. Thus, at best, the surviving text of the beginning of the eruption is a third-hand account. It is also by no means as perceptive as Pliny's letters. Dio Cassius, or his eye witnesses, indulge in flights of fancy that were typical of the period, but which Pliny the Younger eschewed. Dio Cassius claimed that, before the eruption, giants had appeared on the mountain, in the surrounding countryside, and even in the cities near by. They had wandered around both during the day and at night, and had even flitted through the air. The most rational explanation for these fantasies might be that the huge billowing column of fumes that later issued from the volcano could have brought the giants of the mythical past to the minds of the panic-stricken observers.

"Afterwards", Dio Cassius wrote, "fearful droughts and sudden and violent earthquakes took place so that the whole plain [of Campania] seethed, and the summits leaped into the air." These were like the exaggerated portents and prodigies that were often adduced to embellish many natural catastrophes. However, the ground most probably did quake, and it is just possible that "the droughts" had been caused when the local wells dried up. Wells commonly dry up when molten rock rises towards the surface and impedes the circulation of underground waters.

Having got his fanciful portents off his chest, Dio Cassius turned to more credible initial aspects of the eruption. "There were frequent rumblings, some of them were subterranean and resembled thunder; those on the surface were like bellowing; the sea also joined in the roar and the sky re-echoed it." This seems to evoke the start of the eruption when the first noisy explosions reverberated about the sky like a violent thunderstorm. These opening salvoes probably began soon after dawn on 24 August. They formed a plume of fumes and fine ash that could have risen 3 km skywards and distributed a thin layer of very fine pale-grey ash to the east of the volcano, where it is only 3–6 cm thick, for instance, on the remains of two vineyards near Terzigno. It was just enough to encourage curiosity, and then increasing consternation, around the volcano. This ash did not even reach Pompeii.

Much worse was to follow between about 10 a.m. and 11 a.m. The volcano unleashed the main phase of the violent eruption. In a flash, consternation turned to fear. Imaginations ran riot. None of the terrified spectators had ever seen anything remotely like it.

> Then, suddenly [wrote Dio (66.22.4)], a portentous crash was heard, as
> if the mountains were falling down in ruins. First, huge stones were

54

hurled aloft, rising as high as the highest summits. Then came a great quantity of fire and endless smoke, so that the whole atmosphere was obscured and the Sun was entirely hidden, as if it had been eclipsed. Thus, day was turned into night and light into darkness. Some thought the giants were rising again in revolt (for many of their forms could be seen in the fumes . . . and a sound like trumpets was heard). Others believed that the whole universe was falling into chaos or fire. And so they fled; some from their houses into the streets; others from outside into the houses; some from the sea to the land; and others from the land into the sea. In their agitation, they believed that anywhere else would be safer than where they actually were. While this was happening, an inconceivable quantity of ashes was blown out that covered both the land and the sea, and filled all the air. It wrought much and varied injuries upon mankind, their farms and their herds; and destroyed all the fish and birds . . . Furthermore, it buried two entire cities, Herculaneum and Pompeii, the latter place while its populace was seated in the theatre. Indeed, the amount of dust was so great that some of it reached Africa and Syria and Egypt, and also reached Rome, where it filled the air overhead and darkened the Sun.

Misenum: 24 August AD 79, noon

It was hot. Seen from Pliny the Elder's mansion above Misenum, sea, land and sky all merged in a blue haze. To the southeast, Capri and the Sorrento Peninsula were shimmering in the sunshine, their rugged white rocks reflecting the blue of the sea. To the east, the darker blue mass of the Posillipo Peninsula blocked Naples from view. On clear days, the distinctive outline of Somma–Vesuvius could be seen rising 32 km away beyond Posillipo, but on 24 August the heat haze was masking all signs of the mountain. Luckily, a refreshing northwesterly breeze from the Tyrrhenian Sea was cooling the house. All was calm. However, during the past few days, there had been rather more ground tremors than usual around the Bay of Naples – just enough to make people shudder, look up for falling masonry, and then continue their conversations. Nothing very disastrous; nothing broken; no-one had been hurt. The tremors were just a salutary reminder that danger was always lurking underground, even in one of the most beautiful places in the empire.

Pliny the Elder was relaxing. He had spent the morning in the sunshine;

The course of the eruption in AD 79

In the days before the eruption, earthquakes shook Campania, and the shocks probably became more frequent and more clearly centred on the volcano as time went on. They offered a warning that magma was rising within the mountain. In AD 79, the warning went unheeded because no-one then had the slightest idea that volcanoes might give out signs that they were about to erupt. Thus, the eruption took everyone by surprise. Early on 24 August, the magma had risen high enough to encounter the groundwater within the volcano. Masses of steam generated booming explosions that blasted away all the lavas blocking the old volcanic vent, and distributed a thin layer of fine ash to the southeast of the mountain.

As the magma approached the crest of the mountain, the pressures upon it were much reduced. The volcanic gases within it separated out in powerful explosions, which shattered the rising magma into enormous quantities of fumes, hot ash and coarse white pumice that were hurled high into the air. More and more molten rock rose up the vent, more and more explosions shattered it to smithereens, more and more ash and pumice soared skywards. By mid-day, a column like an immense thundercloud was already towering 25 km above the volcano. The dust and ash spread into the stratosphere, while the northwesterly wind blew the coarser fragments of ash and white pumice in an unrelenting deluge well beyond the southeastern flanks of the mountain. Pompeii lay beneath the main axis of this onslaught, and this was where the main accumulation took place. During the afternoon, some 1.40 m of ash and white pumice accumulated in Pompeii. Roofs began to cave in. However, very few fragments fell upon Herculaneum on 24 August, and none at all fell on the area lying up wind to the west and northwest of the little town. Naples, lying even farther to the west, became an obvious place of refuge. Then, after about 8 p.m., slightly different magma began to erupt ash and grey pumice, which, during the next 12 hours, added another 1.30 m to the fragments burying Pompeii. The erupting column reached a height of 32 km.

From 1 a.m. on 25 August, this erupting column of ash and grey pumice collapsed from time to time and generated a series of pyroclastic flows that swept down the flanks of Vesuvius. The first destroyed Herculaneum. The second was larger. It swept not only over Herculaneum again, but spilled northwards and devastated the area within 7 km of the crest of the mountain. The succeeding pyroclastic flows were directed towards the southeast. At 6.30 a.m., the third destroyed Oplontis and reached the northern walls of Pompeii. At 7.30 a.m., the fourth covered the town and killed every living thing within it. At 7.35 a.m. a fifth and even larger pyroclastic flow devastated what remained of Pompeii and almost reached Stabiae. At 8 a.m. on 25 August, the volcano unleashed the sixth and most grandiose pyroclastic flow of all. It surged at least 17 km southwards to Stabiae, where it killed Pliny the Elder, while another branch raced almost 32 km westwards and nearly killed his nephew at Misenum. As the day went on, the eruption began to wane, although darkness still prevailed. A hydrovolcanic phase brought the activity to an end in the early hours of 26 August. And the dawning Sun shone on the exhausted and bereaved survivors.

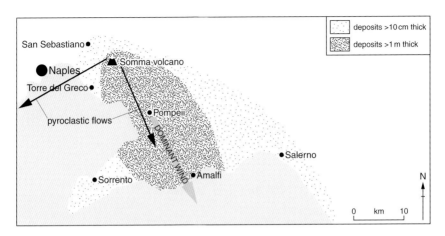

The distribution of airborne fragments and pyroclastic flows erupted in AD 79.

he had refreshed himself with a cold bath, had taken a light lunch while he was lying down, and had started on his books again – a normal sort of holiday morning. An oddly shaped cloud changed all that.

As his nephew later wrote to Cornelius Tacitus:

> At about one o'clock in the afternoon, my mother pointed out to him a cloud that could be seen, which had an odd size and appearance . . . He called for his shoes and climbed up to a place that would give him the best view of the amazing phenomenon. From that distance, it was not clear from which mountain the cloud was rising – although it was afterwards found to be Vesuvius. The cloud could best be described as more like an umbrella pine than any other tree, because it rose high up in a kind of trunk and then divided into branches. I imagine that this was because it was thrust up by a recent blast until its power weakened and it was left unsupported, or was even overwhelmed by its own weight and spread out sideways. Sometimes it looked pale, sometimes it looked mottled or dirty, according to whether it had carried up earth or ash.

Pliny the Younger's description and brief analysis suggests that his uncle had already recognized that the cloud was being erupted from a volcano, similar to those that he had mentioned in his works. However, it is unlikely that the naturalist had ever seen an eruption himself until that morning.

Pliny the Elder did not hesitate. He forgot his weight and his breathlessness, and he seized the chance to take a closer look at this extraordinary feature. "Like a true scholar, my uncle saw at once that it deserved closer study, and he ordered a boat to be prepared. He said that I could go with him if I wished". Most lads of 17 would surely have jumped at the chance to join such an expedition, but he was one of those youths who would rather read a book than sail off to investigate strange clouds on hot afternoons. "I replied," he wrote, "that I would rather continue with my studies. As it happened, he himself had given me some writing to do."

Rectina asks for help

Just as Pliny the Elder was leaving the house, his nephew recounted how "he was handed a message from Rectina, the wife of Tascus, whose home was at the foot of the mountain and she had no way of escape except by boat. She was terrified by the threatening danger and begged him to rescue her from disaster." The surviving texts are clear about neither the spelling of the names nor the exact nature of Rectina and Tascus, but most scholars now agree that Rectina was indeed a lady in distress who was crying for help.

Even so, this is still an amazing anecdote. Rectina probably lived near the lower southeastern flanks of Vesuvius, in the area first affected by the eruption. Her messenger must have travelled by boat rather than on horseback, because Pliny states that this was the only means of escape from Rectina's house. It would take the messenger the best part of three hours to reach Misenum at about, say, 1.30 p.m. Thus, the eruption must have been causing great anxiety by 10 a.m. Rectina must have had extraordinary confidence in Pliny the Elder, who lived some 30 km away across the Bay of Naples, because her messenger could surely have found effective help more easily and more quickly from Naples itself. Alternatively, in her desperation, Rectina had perhaps sent out several messengers to seek help in different directions, and the servant who reached Pliny the Elder was the only one whose journey was recorded. It has even been suggested that Pliny the Younger used a little poetic licence and introduced this incident to heighten the dramatic effect of his uncle's departure on his grand and fatal mission. If Rectina's request was real, it transformed Pliny the Elder from being a scholar who was embarking on a scientific expedition into a hero who set out to save lives without thought for his own safety. As Pliny the Younger emphasized, "What he had started in a spirit of scientific curiosity, he ended as a hero".

Nevertheless, Rectina clearly made a fatal mistake when she stayed at home, waiting to be rescued. She should have dashed for Misenum with her messenger. He reached safety in Misenum; Rectina almost certainly died within the next two days. She had every reason to be terrified; the eruption had already become very violent well before Pliny's sister noticed the cloud at Misenum. The crest of the erupting column was rising higher and higher, and ash was falling over a vast area to the southeast of the volcano. Rectina was soon sharing her terror with thousands of her neighbours – in vain.

Enquiry and rescue: 24 August, afternoon

Before Rectina's messenger had arrived, Pliny the Elder had already asked for a boat to be prepared so that he could embark on his own scientific expedition. This was a liburnica, a light and versatile sailing vessel, like a skiff, that could manoeuvre swiftly. When Pliny the Elder received Rectina's message, he ordered some large and powerful Imperial quadriremes to be prepared to evacuate Rectina and her fellow citizens. These great ships had four ranks of oars that were manned by galley slaves who must have had muscles like titans. The ships would evacuate Rectina's fellow citizens. Unfortunately, however, Pliny the Younger does not record how the evacuation was organized or how successful it proved to be.

Thus, at about 3 p.m., or perhaps a little before, Pliny the Elder set out on his last journey on board the liburnica at the head of an Imperial flotilla of quadriremes; they sailed eastwards out of the harbour at Misenum. They soon left Baiae behind, and the spot where Nero had tried to have his mother, Agrippina, drowned. They passed Pozzuoli and sailed alongside the buff-coloured hump of the island of Nisida and the tree-covered peninsula of Posillipo. The temples and harbour of the city of Naples came into view, but Pliny the Elder ignored the city and aimed for Herculaneum and Oplontis, on the coast at the foot of the volcano. The northwesterly breeze filled the sails of the Pliny's liburnica and kept it skimming ahead of the galleys – making straight for the towering erupting column – and straight into the heart of the greatest danger that Campania had faced for many centuries.

As soon as Pliny had passed Posillipo, he could see that the erupting column was rising many times higher than Vesuvius itself. On its leeward side, a veil of red-tinged grey stretched out like a great rainstorm towards the southeast. It was the ash and pumice that were being winnowed out from the column and blown down wind well beyond Pompeii and Stabiae and over the hills towards Salerno. The terrible roar of the volcanic explosions became ever more threatening, blotting out the swishing of the quadriremes as they sliced through the water, and masking the regular booming from the drums that were beating time on the galleys and forcing the hapless oarsmen to even greater efforts – boom, clunk, splash, swish, sigh, boom, clunk, splash, swish, sigh, boom – on and on, until they arrived almost underneath the fearsome expanding cloud. And they would meet more and more vessels of all shapes and sizes, full to overflowing with desperate people, fleeing, rowing, paddling and screaming, and making their way westwards away from the terrible and overpowering danger. Nevertheless, the naturalist continued on his course, still fearlessly dictating his observations as the catastrophe unfolded.

With a fresh crew, the Imperial quadriremes could travel at a speed of about 15 km an hour in open water, but they would make slower progress when they were handicapped by falling ash and floating pumice as they approached the coast. Pliny's liburnica would probably travel faster. Herculaneum lay about 22 km and Oplontis 28 km in direct line from Misenum. The head of the flotilla would therefore be approaching the coast some two or three hours after it left Misenum. Thus, by about 5 or 6 p.m., Pliny the Elder was probably very close to the shore. Ash and pumice began to fall thick and fast upon the liburnica, which indicates that Pliny the Elder had sailed to a position directly down wind of the volcano, close to Oplontis and beyond Herculaneum, where little ash actually fell until about midnight:

> The ash that was already falling became hotter and thicker as the ships approached the coast. Soon there was pumice and blackened burnt stones that had been shattered by the fire. Suddenly the sea shallowed where the shore was obstructed and choked by debris from the mountain.

It is not clear exactly why the waters shallowed so suddenly. The sea may have retreated, as often happens when volcanic eruptions occur near the sea; but, in this case, the waters would soon return in the form of tsunami waves that crash with devastating violence onto the shore. But Pliny the Younger makes no mention of any such return wave, so that a tsunami can probably be discounted. Alternatively, the sea bed might have been exposed when the land itself rose up, as sometimes occurs when magma surges upwards during eruptions, but again there is no independent evidence of this. The most reasonable explanation – as the words of Pliny the Younger suggest – would seem to be that ash and pumice had suddenly started to rain down in such quantities that the shore had become choked with fragments. Moreover, the pumice would float on the sea and would add, in particular, to the hazards facing the oarsmen. The ships would therefore be able to make further progress only with the greatest difficulty.

Obviously, plans had to be reconsidered. Pliny the Elder stopped, thought for a moment, and consulted his captain. The captain advised him to abandon his attempts at rescue, as well as his scientific quest, and to return to Misenum forthwith. Pliny the Elder disagreed, and thereby made the decision that might have cost him his life. He decided to change tack, take advantage of the northwesterly wind, and go on and await developments at a friend's home in Stabiae, some 12–15 km farther around the Bay of Naples. "Fortune favours the brave", he declared, "take me to Pomponianus!". No-one now knows who Pomponianus was, nor why Pliny the Elder selected his home as a haven, but the liburnica duly set sail for Stabiae.

Unfortunately, Pliny the Younger did not explain what happened to the Imperial galleys that would then be approaching Oplontis. Did their commander leave them in the charge of a subordinate? Were crowds gathered on the shore, clamouring to be rescued? Could any such crowds even be seen, or heard, through the thickening hail of ash and pumice? It is possible that the galleys might have started to rescue the people gathered on the shore; and Pliny the Younger's silence about their activities might therefore merely indicate that he realized that Tacitus would already know what the quadriremes had done on the evening. In any case, these vessels apparently did not follow their august commander to Stabiae.

Pompeii: 24 August, afternoon and evening

In Pompeii, it had been a perfectly normal summer dawn: the hot Sun had risen in a cloudless sky; the northwesterly breeze was just strong enough to temper the heat beating down on the busy streets; business had started as usual. The Pompeiians in the spacious Forum saw an early sign of the eruption and they stared with increasing curiosity at the column of fumes rising above the crest of the mountain. Curiosity turned to anxiety in mid-morning when a vast cloud suddenly broke out from the summit of Vesuvius and soared high into the northern sky before the terrible roar of the explosion reached the astonished spectators. The people turned to each other in apprehension and amazement, seeking someone – anyone – who could reassure them. But in a trice, the gigantic cloud was soaring over the town, and the citizens of Pompeii were fleeing for the shelter of their homes with all the speed that sheer terror could give to their legs.

The explosions were soon following each other so quickly that they merged into a continuous roar. The ground shook. Then, a mixture of fine ash and lumps of white pumice began to fall like dense snow. At first, the falling fragments were cold, but those that followed after an hour or so were warm. Soon they began to rain down on the town like the deluge of some infernal storm. The pumice pattered and then thudded down onto the streets and rooftops. Larger hotter boulders crashed through roofs or smashed to pieces as they hit the streets. The ash swirled into the atriums and into every open room in every house. No-one could breathe normally any more, and each gasp for breath filled their lungs with the finer particles that no end of coughing seemed to clear. It was getting darker and darker as the wind spread out the thickening cloud until it masked the Sun. By the early afternoon, it was pitch dark in Pompeii, and the ash and pumice already lay 15 cm deep.

The Pompeiians were now isolated in small family groups in their homes. It is very doubtful if they realized that they were experiencing a volcanic eruption at uncomfortably close range. They had no real notion of what was going on, and they had no idea what to do. They could neither identify their tormentor nor take any logical steps to combat the dangers that they faced.

The Roman Forum at Pompeii, with Vesuvius on the skyline.

Would it be better to abandon everything and flee? But where should they go? Who knew how far they would have to run to reach safety? Would it be better to stay put in a well sealed room, and keep an eye on the house, or on the business, or on an aged relative? The only thing that seemed clear was that the situation was getting worse with every passing moment. The mountain was roaring ever more loudly, shaking the ground even more vigorously. The choking ash stifled the oil lamps as soon as they could be rekindled. The ash clogged the water clocks and made it impossible to tell the time. Every minute seemed like an hour. The thud of the pumice on the roofs had changed to a dull swishing like hail as the fragments accumulated. Every other sound was muffled as if a thick stultifying fog had enveloped the town. The warm blanket of ash and pumice was thickening in the streets, on the rooftops, and in the open courtyards of every house. It was becoming very hard to open doors. Some unfortunate Pompeiians had no real choice but to stay where they were. In the House of Polybius, for instance, a pregnant woman was nearing her term; her family stayed with her, with father holding the hand of his teenage son. They died together. Then, when the ash and pumice reached a thickness of 60 cm or more, roofs began to cave in.

Now many more citizens opted to gather up their most treasured possessions and try to reach safety. But they had to scramble, in the pitch darkness, through warm ash and pumice that swirled up behind them in blinding and choking clouds with every urgent step that they took. They searched for the gates of the town, but they often did not know which way to run, nor even which way they were running. They ran a desperate race. Nevertheless, many might have succeeded in reaching safety, for relatively few bodies have been found in the streets that were buried *directly* by the falling ash and pumice. However, it is more than likely that many of the fugitives succumbed to exhaustion in the countryside once they had left the town, where their bodies have not yet been excavated. Those who panicked and fled at once probably saved their lives, provided that they had the luck to run southwards away from Vesuvius. Those who took to the boats probably fared best of all, for they would take the safest passage by sailing westwards across the Bay of Naples. Perhaps as many as three quarters of the population of Pompeii took their chance and fled on that black afternoon. Modestus, the baker, was probably one of these. He abandoned his trade and left 81 loaves baking in his oven, where excavators found them overcooked, but still intact, some 1800 years later.

The eruption became even more violent after about 8.00 p.m. on 24

August. Fist-size lumps of grey pumice, mingled with ash, began to shower down upon the town. The fragments soon lay well over 1 m deep in the streets and on the rooftops. Those who hesitated now were already lost. More and more roofs began to creak, snap and fall without warning, cracking the skulls of those sheltering beneath them. The rich mistress of the House of the Faun, one of the most splendid in Pompeii, had collected all her jewellery together. The roof foundered and buried her alive with her treasures. In the House of Trebius Valens, a roof suddenly collapsed on four hapless victims, who had crouched down to avoid being crushed.

The finer ash swirled along the streets, drifting like snow against the doorways, and penetrated into every room, however much the householders had

Victims from Roman Pompeii. When a body decayed, a hollow was left behind in the stiff volcanic ash. The excavators filled such hollows with liquid plaster. When the plaster solidified, it preserved graphic moulds that sometimes revealed folds of clothing and the facial expressions of the dying victims. The victim on the right has been felled by a flying slab in the pyroclastic flow.

tried to seal them. The citizens were stifled by the heat, choked by the ash, and terrified by the muffled roars reverberating from the volcano. They pressed damp cloths over their mouths to filter the air and breathe a little better, as they sat together, shaking, in their innermost rooms. Who knows how many died of suffocation, or even of fright, as that terrible evening wore on? The increasing violence of the eruption probably convinced more and more anguished citizens to abandon their homes and seek safety elsewhere.

Stabiae: 24 August, evening

Pliny the Elder made good speed to Stabiae as the northwesterly wind filled the sails of his liburnica. At Stabiae, the onshore wind made it easy to land but virtually impossible to leave again. Pliny the Elder found that fear had already dictated policy to Pomponianus. When the liburnica landed at about 6 p.m., Pomponianus had left his home and was already installed on the shore where his servants had filled a sailing boat with his more valuable belongings. He had intended to escape as soon as the contrary, onshore, wind changed, but in fact Pliny the Elder's arrival persuaded him to return home:

> My uncle embraced his terrified friend and offered him comfort and encouragement. He believed that he could best calm his friend's fears by demonstrating his own composure, and thus he gave instructions that he was to be taken to the bathroom. After his bath, he lay down and dined. My uncle was quite cheerful, or at least he pretended to be, which was just as brave. Meanwhile, very broad flames, and fires leaping high, blazed out from several places on Vesuvius, and the darkness of the night seemed to make them glare out with even greater brilliance. My uncle soothed the fears of his companions by saying repeatedly that they were nothing more than fires left by the terrified peasants, or empty abandoned houses that were blazing.

Roman heroes obviously knew how to behave in the face of great danger. And the danger was increasing only too clearly. Pliny the Elder took — or affected to take — little heed of this growing threat. He tried to calm the fraught nerves of the lesser mortals around him and he went to bed. After all, it had been a hard day for an old man of 56. Pomponianus and his companions stayed up all night because they were too frightened to go to sleep.

The crisis grew to a crescendo as the night wore on. The ground started to tremble. Every building shook. Ash and pumice fell thicker and faster from the angry sky. The warm fragments were then already lying perhaps 1 m deep in Stabiae. Panic seemed justified and inevitable. If Pliny the Elder was terrified, he did not show it, or at least his nephew claimed that he did not:

> He went to bed and fell properly asleep, because, as he was a stout man, his breathing was loud and heavy and could be heard by those passing his door. But, eventually, the courtyard that gave access to his room began to fill with so much ash and pumice that, if he had stayed in his bedroom, he would never have been able to get out.

Thus, Pliny the Elder spent his last night on Earth.

Herculaneum: 24–25 August

The volcano rose only about 7 km due east of Herculaneum. Throughout the afternoon of 24 August, the inhabitants of the seaside town had a grand-stand view of the eruption. Herculaneum lay up wind of the mountain, so that probably less than 1 cm of falling ash dusted the streets during the whole eruption, and the citizens suffered none of the torment inflicted upon the Pompeiians. Perhaps, too, the breeze carried the sounds away from the town and muted the fierce roaring of the eruption. Nevertheless, as the late after-noon Sun sank into the west, it would only emphasize the fearsome contrasts between the silvery billows of steam and the dark swirls of ash careering sky-wards with such force that the agitated crest of the erupting column towered 33 km above them. Many of the citizens would probably not stay long enough to admire a spectacle that surpassed in horror anything that they had ever set eyes upon. The cloud would be much larger than any thunderstorm, or any mountain, that they had ever witnessed. The citizens would have had no idea what had caused this appalling portent, but they would recognize that it was getting worse all the time. Unlike the Pompeiians, the citizens of Herculaneum could see what they were doing, and they were not being choked by swirling ash when they were doing it. They would see that it would be foolish to flee eastwards, where ash and pumice were clearly rain-ing down on Pompeii and Oplontis. Thus, both calm and panic-stricken citizens would probably react in the same way, and flee in the same direction

– to the west. Westwards lay land that was apparently unaffected by the disturbing events to the east; and westwards, too, lay Naples, the chief city of Campania, where they would be protected. Those who opted to leave had two main escape routes: overland along the coast road, or by boat across the Bay of Naples. Indeed, some of those in the boats in the late afternoon may have caught sight of the Imperial flotilla apparently making for the pumice-choked shore near Oplontis. No doubt they would imagine that madmen had taken control of the vessels.

The excavated ruins of Herculaneum have revealed very few bodies. Perhaps it was soon after nightfall that most of the citizens had abandoned their homes, and there were no stragglers in the streets. Some few had opted to stay behind, or might have been abandoned by callous relatives. A man and a woman stayed together in the men's baths. Two sick men were left in their beds to fend for themselves. A young boy was lying down on his bed in the house of the gem cutter, and a baby was left in its cradle in the House of the Gem. Perhaps the most charitable explanation is that they had already died before their bodies were abandoned.

Many people had already made their way down to the shore, some 15–20 m below the town. Just after midnight, some 300 people were still waiting for boats; they included several cripples and at least a dozen children under the age of three. They all assembled on the beach of black sand, or just

The suburban baths on the waterfront at Herculaneum.

in front of the 12 boat-houses that formed large arches under the splendid new suburban baths at the southwestern edge of the town. No boats were available at that moment. Perhaps everyone was waiting anxiously, but in an orderly Roman fashion, for rescue boats to return and take them to safety in Naples?

The eruption increased in power until after midnight, enlarging the volcanic vent and expelling ever more fragments of molten rock into the stratosphere and onto a vast area to the southeast of the mountain. This sort of activity, now called Plinian activity, is the most powerful in the whole volcanic repertoire, but there are short interludes when the power diminishes and part of the erupting column collapses. Paradoxically, for the people near the volcano these interludes are usually more dangerous than the periods of the most sustained eruptive violence.

At 1 a.m. on 25 August, part of the towering column collapsed like a crumbling pillar of fire. It formed a pyroclastic flow of hot toxic gas, steam, ash and pumice at a temperature approaching 400°C that swept down the western flanks of the mountain at a speed of 250–400 km an hour. Every living thing in its path was doomed. The pyroclastic flow headed straight for Herculaneum. Less than four minutes after it began its lethal journey, it swept into the town, swirled down the streets, and ripped off the rooftops, rushed into every building and penetrated every room.

The base of the pyroclastic flow scythed down the upper storeys of virtually every building in the town. The people waiting anxiously on the shore below the suburban baths heard the sudden dreadful crashing, but they could not have seen the pyroclastic flow itself until the very last moment. In any case, it was already far too late. They had no chance. Sixteen people and a horse died instantly on the beach itself. Their heads exploded, their brains boiled, their teeth cracked, and their bodies shrivelled and dried up at once. Those in front of the suburban baths turned and crowded desperately under the arches, trying to get as far inside the boat-houses as possible to escape from the searing heat. As many as 40 people crushed together in each boat-house. It was of no avail. They had just time to clutch their loved ones. The heat dehydrated their muscles, which flexed their limbs into a boxer's posture that made them seem – wrongly – as if they had died fighting off the surging cloud. Analysis has revealed that they were burned to death, as if they had been struck down and baked by some vast and terrible lightning. Some 1900 years later, excavators found their skeletons in an indescribable tangle betraying a frantic terror that can scarcely be conceived. And the horror of their dying moment was still stamped upon their remains.

Then, the eruption resumed its full power, only for it to falter once again about an hour later. A second, larger and faster-moving, pyroclastic flow rushed down the western slopes of the mountain. It was hotter than the first, with a temperature exceeding 400°C. In Herculaneum, it charred all the exposed wood and corpses, wrecked the surviving upper storeys in the town, and threw the rubble of walls and columns, bricks and tiles onto the shore. It made the sea boil, but there was no living thing left to kill in Herculaneum. Four other pyroclastic flows followed within the next eight hours; they sealed the pathetic remains of Herculaneum under 20 m of fuming ash and pumice, and extended the shore 400 m out to sea. The volcano had claimed its first few hundred victims. No-one knows their names.

Oplontis: 25 August

Oplontis lies beneath part of the present town of Torre Annunziata. It was a small wealthy settlement, about 7 km south of the crest of Vesuvius, that seems to have been composed mostly of villas along the Bay of Naples. One of the most remarkable villas of the Roman world has been excavated there, and its rich wall decorations make it a treasure house of Roman painting. It had probably belonged to Poppaea Sabina, a member of a notable Pompeiian family, who had suffered the misfortune to become the second wife of the Emperor Nero in AD 62. One day in AD 65, she complained when he came home late from a chariot-race meeting. He kicked her; she was pregnant, and she died. But her villa was already unoccupied by then. It seems that it had been severely damaged during the earthquake in AD 62 and, for some unknown reason, repairs had not been undertaken when the eruption occurred. The neglect of such a masterpiece clearly suggests that other palaces were available for occupation in the district, and that perhaps similar magnificent treasures are still awaiting excavation.

The settlement stood close to the main axis of accumulation of the fragments that were being carried southeastwards. Thus, ash and pumice began to rain down on Oplontis almost as soon as the eruption began on the morning of 24 August. As in Pompeii, ash and white pumice piled up quickly at first, but at about 8 p.m. that evening they were followed by falls of ash and grey pumice. The first two westward-moving pyroclastic flows that destroyed Herculaneum apparently missed Oplontis. Then, the volcano began to expel a series of ferocious pyroclastic flows towards the southeast. The first of these

The restored columns and frieze in a portico in the Villa of Poppaea Sabina at Oplontis that had been shattered by the pyroclastic flows.

attacked Oplontis at 6.30 a.m. on 25 August and swirled into the spacious courtyards of the villa, broke many of their surrounding pillars, knocked down the sturdy doors, and buried the beautiful decorations on the walls. Several other pyroclastic flows followed. In fact, more pyroclastic flows buried Oplontis than covered Pompeii, and they made a major contribution to the shroud of fragments, 8 m thick, that masked this treasure house until archaeologists brought it to light in 1964.

Pompeii: 24–25 August: the day of wrath

How long can a human being withstand total terror without going mad? The Pompeiians endured 18 hours of suffering, which was made all the worse because they had no idea about when their distress could possibly end. The column of erupting gas, steam and volcanic fragments of all sizes was soaring 33 km skywards. The ash and pumice were raining down thicker and faster on Pompeii, accumulating so thickly that many ground-floor doors would no longer open. More and more roofs were collapsing under the weight of the fragments, crushing the hapless citizens trying to shelter beneath them. Whenever people moved, they sent up swirling clouds of dust and ash. Domestic animals howled and whimpered. The very young and the elderly were struggling to breathe. Everyone was thirsty. Death hovered so closely above the people of Pompeii that the night must have seemed an eternity. But death would strike them before they would see another dawn – in an even more horrifying manner than any of them could ever have imagined.

As in Oplontis, the pyroclastic flows did not trouble Pompeii during the night of 24–25 August, but the inexorable accumulation of fallen ash and pumice persuaded some of those who could still master a few rational thoughts to try and leave the town. It was now more easily said than done. By 6 a.m. on 25 August, the warm fragments were lying over 2.8 m deep in the streets, and the ground floors of many buildings were completely buried. The lightning and the glows flashing from the erupting ash provided the only glimmer of light to guide the fugitives on their journey.

At 6.30 a.m. on 25 August, the same pyroclastic flow that had crashed over Oplontis raced on southeastwards – directly towards Pompeii, but it lost its impetus and halted just at the foot of the northern walls of Pompeii. It would have seemed to an impartial observer that the volcano was tormenting Pompeii, but there were no impartial observers. Few of the citizens could even see the faces of their families quivering beside them. Suddenly the dangers seemed to multiply. Terrifying earthquakes began to shake every wall and floor in the town, and each shock sent up vast clouds of dust and ash that almost suffocated the people still sheltering in their homes.

The addition of the earthquakes to the seemingly infinite catalogue of threats menacing the Pompeiians might have induced a further few hundred people to leave the increasingly fragile shelter of their homes and to take their chance in the appalling atmosphere outside. They set off, clutching their most treasured possessions, not knowing where to run. Not one man, woman or child reached their destinations.

At 7.30 a.m., a fourth and more powerful pyroclastic flow swept into Pompeii. It knocked flat most of the upper storeys of the buildings, tore off the remaining roofs, smashed down many walls and added their masonry to its armoury of red-hot ash, pumice and poisonous gas. No human being exposed to such an onslaught could survive. Perhaps more than 2000 people still remained in the town. Many suffocated, some were burned to death, some were battered to death by lumps of masonry, some were baked, a few might have been gassed. No matter what killed them, their meagre consolation was that they would suffer for no more than about two minutes before death released them from their agony.

Those struggling through the ash in the streets might have had just enough time to turn and run when they saw the glowing cloud rushing down upon them. The pyroclastic flow threw them to the ground with a disdainful swipe. The searing ash in the flow choked them and burned out their lungs in the space of a few breaths as they tried desperately seal their mouths. They had to breathe to live; if they breathed they would die. At the Nola gate, on the northeastern walls of the town, the pyroclastic flow scythed down 33 people who were, in fact, fleeing straight towards it. Some were carrying talismans, the keys to their homes, or a few coins in a little purse, and two victims were carrying oil lamps in a poignant attempt to light their way. Near the harbour, dozens of people were fleeing in the right direction with their cash, jewellery and silverware. When the danger threatened, they had just time to shelter in a row of 20 shops along the road southwards to Stabiae, but the flow killed them nonetheless. The victims who died near the Nocera gate, on the southern walls of the town, were also at least running in the right direction, away from the pyroclastic flow. A slave, bent double under a great sack of provisions, fell down dead. An old man stumbled and died as he tried to raise himself onto his elbows and push his mouth away from the stifling hot ash so that he could breathe. A woman, perhaps his daughter, died just in front of him as she stuffed her handkerchief into her mouth in a vain attempt to stop the hot ash from entering her lungs.

In the Forum, the pyroclastic flow smashed the colonnade and crushed two priests. It asphyxiated another priest as he ran down the main street, trying to carry away the treasury from the Temple of Isis. And the pyroclastic flow smothered and baked two small boys, holding hands, as they tried to shelter in a corner beneath a roof tile. Most of the Pompeiians who perished in the town at that moment were granted the modest privilege of dying in their own homes. Many families had gathered together in one room, where perhaps they thought that they would be safest, trying in vain to keep out the noxious ash and fumes. The four members of the family in the Diomedes villa, for instance, had barricaded themselves in one room with all their jewellery and valuables, 10 gold pieces, 88 pieces of silver, and with enough food to last through the crisis. Then the head of the family changed his mind and decided to abandon the villa. Key in hand, he was just leading his family out of the house when the pyroclastic flow struck them down. The pyroclastic flow killed a group of nine men and two slaves, guided by a porter carrying a lamp, as they were leaving the House of Menander. They must have decided to leave on the spur of the moment, because they had left behind 118 items of silverware, weighing 24 kg, hidden in a box in the bathroom. They had taken pillows to protect their heads against the battering ash and pumice. Pillows are useless against a pyroclastic flow; they might just as well have raised a white flag.

In the palestra, the exercise courtyard of the gymnasium, a surgeon fell down with his bag of surgical instruments beside him; and an athlete died, still clutching his bottle of body oil. The 60 gladiators in the town had decided to stay put, and they all expired together in their barracks. Two of them had no choice: they died in their chains. Sad to say, the oft-repeated tale that one gladiator died in the arms of an emerald-decked lover has proved to be no more than a sentimental rumour. It seems that the lady had stopped in the barracks only to take shelter from the falling ash and pumice as she and her 18 companions were rushing from the town. Household pets suffered as much as their owners. In one house, a chained dog died as it

This picture, taken in 1902 in Saint-Pierre, Martinique, illustrates the likely fate of many people when the pyroclastic flows struck Pompeii. The remains of the victims are mixed with the ash, rocks and masonry in the flows that killed them.

73

arched its back in one last desperate effort to break loose; another dog in another house was eating its dead master when the pyroclastic flow struck.

By 7.35 a.m. no-one was left alive in Pompeii. The fifth and sixth pyroclastic flows were even larger than their predecessors and they swept over the remains of Pompeii and reached as far as Stabiae. They smashed down the remaining upper storeys of the taller buildings in Pompeii. Then, in the darkness throughout the day, first pumice and ash, and then ash alone, continued to rain gently down, until little trace of the town emerged from its shroud.

Stabiae: 25 August, dawn

Stabiae lay about 17 km from the crest of the volcano, where the effects of the eruption were much less severe than at the foot of the mountain. Driving ash and pumice masked the mountain from sight, but, whenever the clouds of ash and pumice parted from time to time, the flashing lightning and the glow from the molten rocks being hurled out only increased the alarm in the mind of every observer. At about 6.30 a.m., the ground had also started to quake in a terrifying manner. It seemed only a matter of time before Stabiae would be swallowed into Hades.

Pomponianus and his companions eventually became so afraid that they felt they had to waken Pliny the Elder and alert him to the danger. Perhaps they also hoped that the great man might possibly find some solution to their predicament – or would at least be able to tell them what to do:

> He was awakened, and he joined Pomponianus and his servants, who had sat up all night. They wondered whether to stay indoors or go out and take their chance in the open, because the buildings were shaking with frequent violent tremors and appeared to be swaying backwards and forwards as though uprooted from their foundations. Outside, there was the danger from the falling pumice, although it was only light and porous. After weighing up the risks, they opted for the open country. For my uncle, it was a choice between rational considerations, for the others it was a choice between fears. To protect themselves from falling objects, they tied pillows over their heads with cloths.

Pomponianus and his companions scampered out into the open country in the first direction that entered their minds. With undignified haste, they ran out of history with pillows on their heads, and were never heard of again. Dawn should have broken by then, but it was still as black as night in Stabiae. Nevertheless, Pliny the Elder kept his composure as the crisis gained momentum. He went down to the shore with a couple of slaves to assess the situation. As his nephew recounted, Pliny the Elder was rational to the last:

> It was daylight everywhere else by this time, but they were still enveloped in a darkness that was blacker and denser than any night, but it was relieved by many torches and various other lights. The decision was taken to go down to the shore to see at close quarters if there was any chance of escape by sea, but the current was still dangerous and running into the shore. He lay down to rest on a sheet that had been spread on the ground for him and he called once or twice for a drink of cold water and gulped it down. Then, suddenly, flames and a strong smell of sulphur that warned of yet more flames to come, forced the others to flee and him to get up. He stood up, with the support of two slaves, and then he suddenly collapsed. I imagine that it was because he was suffocated when the dense fumes choked him by blocking his windpipe, which was constitutionally weak, narrow, and often inflamed.

Thus it was that, on the shore at Stabiae, at about 8.00 a.m., the volcano claimed its most distinguished victim. Its sixth and largest pyroclastic flow had spread 17 km southeastwards from the crater. Even at such a distance, the pyroclastic flow would still be laden with a whole gamut of lethal weapons, ranging from masonry to hot ash. A man of Pliny the Elder's rather precarious health would have no chance of survival. It would merely be a question of which particular weapon in the pyroclastic flow would strike him down.

In fact, the exact cause of Pliny the Elder's dramatic death is uncertain. Some have agreed with his nephew's view that he had been suffocated by dense fumes. Some have suggested that his death was brought on by a severe asthmatic attack. Some have believed that he was asphyxiated by a toxic gas, such as carbon dioxide or hydrogen sulphide. Others have argued that he could have died from a heart attack brought on by his repeated exertions in the stifling atmosphere. Suetonius even claimed that Pliny the Elder had either been overcome by the dust and ash, or had been unable to withstand the suffocating heat any more, and had therefore commanded one of his slaves to hasten his end. Unlike most victims of pyroclastic flows, Pliny the Elder did not die instantly. When he had gone down to the shore, he had already been forced to lie down on a sheet and had twice gulped down water. He had already been taken ill when the smell of sulphur forced his companions to flee. Did the gases poison him or merely accelerate his end?

Two slaves were with Pliny the Elder at the time. His nephew, of course, made no mention of their fate. This is a regrettable omission. If the slaves survived, then the chances are that Pliny the Elder died from natural causes, albeit perhaps accelerated by his exertions. On the other hand, if the slaves died with him, then the pyroclastic flow must surely have killed all three. The exact role of the pyroclastic flow would, of course, still remain to be determined. Pliny the Younger might also have gleaned more information from Pomponianus or his companions, or, indeed, from the crew of the liburnica, whose fate is not recorded, although they could have returned to Misenum. On the other hand, at the time, all these likely witnesses would probably have been too busy saving their own lives to stop and watch Pliny the Elder die.

After more than 1900 years, who can tell exactly what happened? So many deaths in antiquity are shrouded in mystery that it seems far too much to expect a clear analysis of a famous scholar's death – especially in the midst of one of the most violent volcanic eruptions of the age. In any case, there is an air of "official history" about Pliny the Younger's account of the sequel to his uncle's death. He later declared that his uncle's body had been found intact and uncorrupted when the eruption drew to a close on 26 August. However, his statement is not as helpful as it might seem, because it could have been designed to emphasize that his uncle had died a distinguished death. From this life, he had made the worthy departure of a great man and he seemed to be in the eternal sleep of a hero.

Misenum: 24–25 August

At the end of his first letter to Cornelius Tacitus, Pliny the Younger added a postscript, perhaps to tempt his friend to ask for more information about the course of events that he had witnessed at Misenum. If that was his aim, he certainly succeeded. Tacitus did indeed ask for more, and history has every reason to be grateful for a vivid description of the impact of a great eruption on a panic-stricken population.

Pliny the Younger's account of the events at Misenum was published as Letter 20 in Book 6 of his *Letters*. It was a story of despair and horror that clearly reminded him of the last night of Troy, which Virgil described in the

Aeneid. Indeed, he started his second letter by quoting the famous line 12 from Book 2: "Although my mind shrinks from remembering and recoils in the face of so many bereavements, I will begin". It seemed like the end of the world. And for some it was:

> After my uncle left, I spent the rest of the day at my books, because this had been the reason I had stayed behind . . . We had experienced earth tremors for several days, which were not especially alarming in themselves, because they happen so often in Campania. But that night they were so violent that everything felt as if it were not only being shaken but also destroyed. My mother came hurrying to my room and found me already getting up to wake her if she had still been asleep. We sat together in the forecourt between the buildings and the sea near by. I do not know whether I should call this courage or foolishness on my part (I was only 17 at the time), but I asked for a volume of Livy and went on reading as if I had nothing else to do. I even continued with the extracts that I had been making. A friend of my uncle's had just arrived from Spain to join him. When he came up and saw us sitting there, and me reading indeed, he scolded both of us – me for my folly, and my mother for permitting it. Nevertheless, I remained intent upon my book.

As the night progressed, the increasing violence of the eruption was just as evident at Misenum as it was at Stabiae at the other side of the bay. No ash had fallen on Misenum, because it was well up wind of the erupting crater, but the column had risen into the stratosphere and had spread over a vast area to the east of the naval base. Its pall of fumes and dust masked the rising Sun from view at Misenum. This veil was sinister, but it was by no means as terrifying as the increasingly violent earthquakes that were now shaking the whole district. The land seemed to be in constant motion.

> By now it was past six o'clock, and the dawn light was still only dim and feeble. The buildings around were already tottering and we would have been in immediate danger if our house had collapsed, because we were in such a small and confined space. This made us decide to leave town. We were followed and pressed ever forward by a dense and panic-stricken crowd that chose to follow someone else's judgement rather than decide anything for themselves – a point when fear looks like prudence. We stopped once we were beyond the buildings of

the town [probably on the hill of Monte di Pròcida that overlooked Misenum], when some extraordinary and most alarming things happened. Although the ground was flat, the carriages that we had ordered began to lurch to and fro and we could not keep them still – even when we wedged their wheels with stones.

The sky was dark, the Sun was masked, and the land was in turmoil. Now it was the turn of the sea. Below them, the astonished crowd of fugitives "saw the sea sucked away, and then apparently forced back by an earthquake. Certainly, the shoreline had retreated and many sea creatures were left stranded on the dry sand." This was probably the first stage of a tsunami, but Pliny the Younger's attention was soon diverted by an even more appalling threat.

It was about 8.a.m. when the sixth and greatest pyroclastic flow erupted. Unbeknown to Pliny the Younger, it killed his uncle at Stabiae. It also came within a few metres of exterminating him, his mother and the crowd of hapless fugitives congregated on the Monte di Pròcida. Pliny the Younger seems to have recalled Virgil's vivid description of the apocalyptic last night of Troy in Book 2 of the *Aeneid*. When the Greeks had sacked the city, Aeneas had put his old blind father onto his shoulders and had taken his own son, Ascanius, by the hand and had led them both to safety. Thus, Aeneas had begun his epoch-making journey to lay the foundations of Rome: Pliny the Younger led his mother by the hand from Misenum to reach safety on Monte di Pròcida. The symbolic parallel was perhaps exaggerated, but the anguish was the same. It is difficult to say whether the aspect of a pyroclastic flow near Misenum, 32 km from its source, would be more dangerous and terrifying than the sight of Troy being torched and put to the sword. Given the choice, most modern volcanologists would almost certainly prefer to take their chance against the Greeks in Troy:

From the other direction, a dreadful black cloud, torn by whirling, quivering bursts of fiery gas, parted to reveal long tongues of flames like lightning, but larger. At this point, my uncle's friend from Spain spoke out even more urgently. "If your brother, if your uncle is still alive, he would want both of you to be saved. If he is dead, he wanted both of you to survive him – so why postpone your escape?" We replied that we would not think of considering our own safety as long as we could not be certain about his. Without further ado, our friend rushed off and hurried out of danger as fast as he could.

Pliny the Younger is perhaps implying that the Spanish friend, like Pomponianus at Stabiae, lacked the moral strength of a true Roman. Both panicked and fled to save their own skins, and ran into deserved oblivion.

The cloud sank down onto the ground soon afterwards and covered the sea. It had already surrounded and hidden Capri from sight, and had concealed the projecting part of [Cape] Misenum. My mother begged, implored, and then commanded me to leave her and escape as best I could. A young man [she said] might be able to escape, whereas she was old and slow. She would be able to die in peace provided that she had not been the cause of my death as well.

In the *Aeneid*, an elderly relative had also handicapped a vigorous son during the sack of Troy. At first, Anchises had refused to flee from Troy and go to a nearby mountain, because he was weak and old; and the sentiments that he had uttered to Aeneas were very similar to those Plinia had expressed to her son. Anchises eventually allowed himself to be persuaded to escape from the city on his son's shoulders, as the flames from Troy approached them through the thick darkness. However, Plinia did not force her son to go to such extremes:

I told her that I refused to save myself without her. I took her hand and made her hurry along with me. She agreed reluctantly and blamed herself for slowing me down. Ash was already falling by now, but not yet very thickly. Then I turned around: a thick black cloud threatened us from behind, pursuing us and spreading over the land like a flood. [This was the front of the sixth and largest pyroclastic flow from Vesuvius.] "Let us leave the road while we can still see", I said, "or we will be knocked down and trampled by the crowd in the darkness". We had hardly sat down to rest when the darkness spread over us. But it was not the darkness of a moonless or cloudy night, but just as if the lamp had been put out in a completely closed room.

You could have heard women shrieking, children crying and men shouting. Some were calling for their parents, some for their children, some for their wives, and trying to recognize them by their voices. These people were bewailing their own fate, or those of their relatives. Some people were so frightened of dying that they actually prayed for death. Many begged for the help of the gods, but even more imagined that there were no gods left and that the last eternal night had fallen on the world.

The citizens of Misenum were no less than 32 km from the volcano, but horror had taken charge of their minds so completely that they had abandoned all hope of salvation. It was as if they were gathered at the gates of the Underworld. What, then, can have been the anguish of the Pompeiians when they had been faced with an altogether more acute and pressing terror? And in Misenum, too, panic effectively detached brains from mouths and generated rumours of even more menacing threats:

There were also those who added to our real perils by inventing fictitious dangers. Some claimed that part of Misenum had collapsed or that another part was on fire. It was untrue, but they could always find someone to believe them.

At length, the fearsome dangers from the pyroclastic flows seemed to be waning. Ash began to fall heavily, but this was more inconvenient than life threatening. The eruption had passed its climax:

A glimmer of light returned, but we took this to be a warning of approaching fire rather than daylight. But the fires stayed some distance away. [This might have been another pyroclastic flow.] The darkness came back and ash began to fall again, this time in heavier showers. We had to get up from time to time to shake it off, or we would have been crushed and buried under its weight.

Pliny the Younger did not let himself indulge in any false hopes. He believed that he was not only witnessing, but actually taking part in, the end of the world. This was even more grandiose than the fall of Troy. Their plight was too grave for weeping and wailing. Pliny took consolation that others were joining him as they faced up to the ultimate and most solitary act of life – death itself:

I could boast that I never expressed any groan or cry of fear in this time of peril, but I was kept going only by the thought that I was perishing together with everyone else, and that everything was perishing with me, a wretched, but considerable, consolation for my mortal condition.

Death was not to strike Pliny the Younger just yet. The eruption waned; the Sun shone. What it revealed scarcely allayed the fears or calmed the nerves of the survivors. As far as the eye could see, pale ash and pumice was enshrouding the whole landscape. The fugitives kicked up clouds of dust as they made their way back home to await news of their relatives and friends:

> After a while, the darkness paled into smoke or mist, and the real day-light returned, but the Sun shone as wanly as it does during an eclipse. We were terrified by what we saw, because everything had changed and was buried deep in ash like snow. We went back to Misenum, where we attended to our physical needs as best we could, and then spent an anxious night [25–26 August] switching between hope and fear. Fear was uppermost because the earth tremors were continuing and several hysterical people made their own and other people's misfortunes seem trivial with their terrifying predictions. But even then, in spite of the dangers we had undergone and were still expecting, my mother and I still had no intention of leaving until we received news of my uncle.

Victims of the eruption

On the first day of the eruption, it was much easier to escape from Herculaneum than from Pompeii; thus, most probably, a greater proportion of the population survived from Herculaneum. In Pompeii, the remains of about 1150 bodies have been found. Since the total population has been estimated at some 20 000, the bare figures would suggest that only just over 5 per cent of the population of the town fell victim to the eruption. This figure is almost certainly far too low. For instance, from recent studies of the distribution of the remains already excavated, it has been estimated that about 500 bodies could still be buried in the unexcavated parts of the town, and many more bodies probably lie in the vast area around the town that was enshrouded by the ash and pumice. Those who panicked first and fled would have by far the best chance of survival, but it is impossible even to guess how many of these could have escaped. Falling ash and pumice killed those who died during the early hours of the eruption, but as soon as the pyroclastic flows were unleashed, they became the main agents of death. Very few fit men have been found among the victims in Pompeii – probably because they could run away quickly, although they could never outstrip the pyroclastic flows, which would simply overtake them beyond the walls of the town. The majority of the victims were found in the central parts of Pompeii, and very few bodies have come to light in the peripheral districts, except in the exercise court-yard, the palestra, and along the cemetery road leading beyond the north-western walls of the town. Perhaps those living nearer the town walls had already escaped, because they could more easily find their way to the gates in the darkness and the driving ash and pumice? Perhaps, they too, like their fitter fellow citizens, met their deaths beyond the town walls and have yet to be discovered?

It is very unlikely that anyone survived the pyroclastic flows except on their outermost edges. At Stabiae, for instance, at least one person from the household of Pomponianus or the crew of the liburnica apparently lived to tell Pliny the Younger the tale of his uncle's death. On the other hand, there is no means of guessing how many people survived the pyroclastic flow that could have enveloped Naples as it swept across the bay and almost reached Misenum. However, one incontrovertible fact remains: not a single person from Pompeii itself is *known* to have survived the eruption.

Aftermath

When the eruption ended, Somma–Vesuvius was a wreck. An area of $300\,km^2$ around the volcano was completely devastated and buried in a blanket of fragments between $5\,km^3$ and $10\,km^3$ in volume. In that whole area, not a living thing was to be seen. Towns, ports, villas, houses, trees, vineyards, vegetable plots and thousands of people had vanished from the face of the Earth. A sprinkling of windblown ash had fallen on Rome, but also much farther afield on North Africa, and even on distant Egypt and Syria.

The caldera on the summit formed a broad ellipse, stretching about 5 km from east to west and 3.5 km from north to south. The lowest sector of the caldera rim occurred on its southern and southeastern side, which means that lava flows or pyroclastic flows are almost always directed towards the vulnerable and populous towns to the south of the volcano. However, all the flanks of the volcano could be covered by ash and pumice thrown out by more violent eruptions. It is the dominant wind at the time of the eruption that determines their distribution. The dominant westerly winds usually take these fragments to the east. It was the misfortune of Pompeii that a northwesterly wind was blowing at the end of August AD 79 and that the town lay in the very centre of the axis of accumulation of the volcanic fragments.

Aid

When the extent of the disaster became known in Rome, Emperor Titus remained true to his high principles. He started organizing help for the stricken area immediately and he even took the most unusual step of leaving Rome and going to direct relief operations in Campania. Thus, Dio Cassius reported that Titus was absent in Campania when the great fire broke out in Rome in AD 80. (Clearly, Titus did not have the luck that his high principles merited.) Both Dio Cassius and Suetonius recorded that, after the emperor returned to Rome, he nominated two distinguished former consuls and sent them to Campania at the head of what amounted to a colonial expedition to supervise the restoration of the region. Titus used many available public funds to restore all the regions damaged in the eruption, and gave money out of his own pocket to the survivors who were suffering hardship. He even refused to accept the large legacies that rich Romans commonly bequeathed to the emperor (in the usually vain hope that the emperor would perhaps be less tempted to despoil their heirs). Instead, Titus diverted these considerable sums to those in need in Campania.

Unfortunately, no descriptions have survived of the kinds of public works that were undertaken at the time. One of the first priorities of the Emperor Titus must have been to re-establish the Imperial road network throughout the region. Thus, for instance, the old road from Nocera to Stabiae was swept clear and bordered with a ramp of rock to prevent the ash in the fields from being blown over it again. But much of the area remained in ruins for a long time. The towns lying close to the zone of maximum ash and pumice accumulation had almost vanished from the face of the Earth. A thin cover of newly erupted ash commonly weathers down fairly quickly and actually increases the fertility of the soil; but, in a vast area to the southeast of the volcano, the fragments accumulated so thickly that all signs of agriculture were obliterated. The cover of ash and pumice was not only deep but also porous, parched and unable to retain moisture. Regeneration was stifled for decades and it was several centuries before the blanket of volcanic fragments weathered into fertile soils that could restore agricultural wealth to the region. And old Pompeii and Herculaneum were nowhere to be seen.

In fact, there would have been very little that Imperial initiatives could have done about much of the devastation. Indeed, an Imperial tragedy compounded the catastrophe. In September AD 81, Emperor Titus was staying at the spa of Aquae Cutiliae (Rieti), where Vespasian had died just over two years earlier. There, the emperor, a fit man of 41, suddenly fell seriously ill. His brother, Domitian, who was cast in a different mould, saw the chance to seize the power for which he craved. Domitian insisted that Titus could be cured only by a severe chilling, and had the powerless Titus placed in a chest packed with snow. While Titus was breathing his last, Domitian was already making post-haste for Rome. Once in the capital, he distributed a bounty to the Imperial soldiers, which was liberal enough to buy their scruples, so that they had no qualms about proclaiming him emperor forthwith. They thereby instituted a reign of terror that was to last for 15 years. It is not explicitly stated, but it is unlikely that the psychopathic Domitian had time for public works and regional regeneration in Campania.

The effects of the catastrophe naturally attracted the attention of the poets. In AD 88, for example, the Spanish poet, Martial, mentioned the devastation in Campania in his *Epigrams* (4.44):

This is Vesuvius, but lately green with the shade of vines. Here the noble grape loaded the vats to overflowing . . . This was the dwelling of Venus . . . All lies sunk in flames and ashes drear. Those themselves on high would rather this had not been in their power.

In AD 95, the Neapolitan poet, Statius, recorded in Book 4 of his *Silvae* that Vesuvius had covered and condemned a vast area and he indicated that the volcano was still erupting in about AD 90:

. . . where Vesuvius rears his broken wrath, rolling out fires to rival Trinacrian flames [Etna in Sicily] . . . Shall the future progeny of man believe, when crops grow again and this desert shall once more be green, that cities and peoples are buried below, and that an ancestral countryside vanished in a common doom? Nor does the summit yet cease its deadly threat . . .

A more favourable situation prevailed around the margins of the devastated regions. Town such as Cumae, Pozzuoli, Baiae, Misenum and Capua were soon flourishing almost as much as before the disaster – but these places had been far from the cataclysm. So, too, were the islands of Ischia, Capri and Pròcida, and the western end of the Sorrento Peninsula. In Stabiae, the eruption had not completely emptied the town, and many of those who had fled in terror eventually plucked up the courage to return to what was left of their homes. Thus, after a few decades, the population began to rise again; and a Christian community, for example, seems to have been living in Stabiae in the third century.

The eruption seems to have had economic repercussions on Rome itself, although no documents have apparently come to light that expressly blame the volcano. At about this time, the capital began to import increasing quantities of wine, olive oil and fish, most notably from Gaul. The most plausible explanation would seem to be that attempts were being made to compensate for the shortages caused when the rich lands around the volcano had been rendered sterile and the fishing fleet severely depleted. Needless to say, the poets and the aristocracy steered clear of Campania for decades. Hadrian was the first emperor to return to stay in the district, and that was some 60 years after the cataclysm.

Further reading

Cioni et al. 1999; Gigante 1982; Giubelli (undated); Hine 2002; Jashemski & Meyer 2002; Pliny the Younger; Renna 1992; Sigurdsson 1982, 1985, 2002.

Chapter 5

From antiquity to the Renaissance: tall stories

References to Medieval eruptions must be treated with great caution, because they are not only sparse, brief and uncritical, but also commonly little more than asides in tales of the lives of kings and saints.

Somma–Vesuvius did not stay dormant for very long between the zenith of the Roman Empire in about AD 120 and the High Renaissance some 1400 years later. Although one of these eruptions, at about the end of the fifth century, was violent and destructive, many of the remainder were mainly constructive and only moderately explosive. Most of the activity seems to have been concentrated between about 787 and 1139, and it produced ash, cinders and lava flows. As a result, by the time the Middle Ages reached their climax, the cone of Vesuvius had grown into a substantial volcano in its own right, rising more than 700 m above the old caldera floor, which had last been modified in AD 79. After 1139, the volcano rested for several centuries. Activity moved for a while to the Campi Flegrei, where one outburst spluttered up mud within the crater of La Solfatara for a day or two in 1198, and an eruption lasted for a couple of months in 1302 on the island of Ischia.

Limitation of sources

Throughout the period, one of the main volcanological problems is to discover when the more moderate effusive–explosive activity began to dominate the behaviour of the volcano, because these moderate explosions built up the cone of modern Vesuvius within the old summit caldera; and the effusions gave off many lava flows, which had not featured in the volcano's activities for perhaps a thousand years or more. There is also a problem with the

Major eruptions of Somma–Vesuvius, AD 79–1500

AD 79 24–26 Aug.	Large Plinian eruption, with towering eruptive column, ash and pumice falls and pyroclastic flows. Herculaneum, Pompeii, Oplontis and Stabiae buried (2000+ deaths) (P) (VEI 5–6)
AD 80 – c. AD 203	Strombolian activity (S)
AD 203	Strong explosions (SP?) (VEI 3–4?)
AD 203–35	Persistent Strombolian activity (S)
AD 379–95	Possible Strombolian activity? (S)
c. AD 472 5–6 Nov.	Pollena eruption. Powerful eruption (SP) (VEI 4?); possibly the same as in c. AD 507–512.
c. AD 507–512	Powerful eruption of pumice and ash (VEI 4?); possibly the same as the eruption in c. AD 472.
AD 536	Possible Strombolian activity? (S)
685 Feb.–March	Explosive eruption with pyroclastic flows to the sea (VEI 3?)
787 Oct.–Dec.	Lava fountains and long lava flows (VEI 2–3?)
c. 850–1037	Persistent Strombolian activity (S)
968	Lava fountains and lava flows reaching the sea (VEI 2–3?)
c. 991 – c. 993	Strong explosive eruption; possibly the same as in 1007 (VEI 3?)
999	Strong eruption of ash, cinders and lava flows (VEI 3?)
1007	Strong explosive eruption; possibly the same as in c. 991–3 (VEI 3?)
1037 27 Jan.–Feb.	Extensive lava flows reaching the sea (VEI 2–3?)
1068–1078	Possible explosions of widespread ash? (VEI 2–3?)
1139 1–29 June	Strong explosive eruption and lava flows (VEI 3?)
1500	Possible very brief explosion? (VEI 2?)

VEI = volcanic explosivity index (devised to compare the power of different eruptions)   S = persistent moderate effusive–explosive activity of Strombolian type   SP = sub-Plinian eruption
P = Plinian eruption
N.B. The types of eruption are not as clearly distinct as this table implies.

sources of information, which handicaps all medieval references to the natural landscape because reports of eruptions anywhere in the world in the Middle Ages have to be taken with a pinch of salt or a whiff of sulphur. They offer flimsy accounts of volcanic activity: scanty, vague, brief, usually fanciful or exaggerated, and quite often muddled and ambiguous. They occur almost as if by chance, as asides in chronicles devoted to prodigies, accounts of military campaigns, the lives of saints, and occasional reports of interventions of

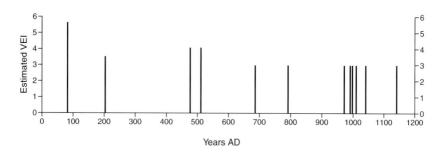

Generalized diagram of the chief recorded eruptions of Somma–Vesuvius from AD 79 to 1139.

the Devil. It is often not even clear when the eruptions took place, how many there were, and, still less, how long they lasted. Most of the time, the hard evidence of eruptions is usually less convincing than many fairy tales.

The medieval chroniclers had no training in volcanological studies, and they did not even have the words to express what they had seen. They neither aimed nor claimed to be scientists studying the Earth; the subject did not exist. The concept of a volcano as a special sort of mountain did not exist. Even the word "volcano" did not exist. Very few people had the slightest interest in scientific or rational explanations of natural phenomena; the revealed truth was enough. Everyone knew the cause of these acts of nature: it was God. Sometimes God rewarded deserving people, at other times, divine retribution punished sin with such catastrophes as volcanic eruptions. The Earth sciences were stunted still further because most of the literate population belonged to the religious orders, who had a different agenda, devoted to the glory of God. Independent thinking, too, was likely to breed heresy, and to upset the established social order into the bargain. The Earth sciences were relegated to an intellectual backwater, understood no better than the propagation of the plague.

Consequently, any account of the medieval volcanic eruptions in Campania is even more speculative than in Roman times, or during any other subsequent period. In fact, only one surviving account shows any attempt to ask the most general of scientific questions and, even then, the specific date of that eruption remains uncertain.

A legend

The writer Matilde Serao recounted one of the tales about the origin of Vesuvius. It has a typical Neapolitan flavour.

Once upon a time there lived in Naples a young knight called Vesuvio who fell hopelessly in love with a young girl called Capri. Faced with the opposition of her family, the young girl threw herself into the sea, from which arose the blue and verdant island that bears her name. When the young knight learned the dire news, he started to give out hot sighs and tears of fire as a sign of the passion raging within him. He was so agitated that he became a mountain in whose entrails burned the eternal fires of love. Thus, he stands opposite his lovely Capri, but he cannot reach her; and he trembles with love, flashes out lightning and crowns himself with smoke; and the fire spills out in gleaming lava.

Eruptions from AD 79 until 685

For six centuries after the great catastrophe of AD 79, Vesuvius erupted infrequently, but, although its efforts were often quite vigorous, none approached the scale of the great outburst in AD 79. They produced much ash and pumice, perhaps several pyroclastic flows, but very few lava flows, if any at all.

An eruption in about AD 170?

Brief references in Martial and Statius suggest that some kind of mild and fairly persistent activity probably occurred on the volcano between AD 80 and AD 120. In about AD 170, Galen, the Greek physician to the Emperor Marcus Aurelius, noted that Vesuvius was famous for the fire that was produced from within it:

A large quantity of ash reaches the sea, which is the remains of the material that has been burning and is still burning [in the crater?], and all this makes the air dry.

Galen's comment could suggest some activity at that time, but unfortunately it is not clear how long such an eruption could have lasted, or even exactly when it happened. Similar forms of mild activity might also have occurred on and off during the next 60 years.

The eruption in AD 203

In AD 203, during the reign of the Emperor Septimius Severus, the volcano erupted rather more violently. In that year, the historian, Dio Cassius, who was staying in the town of Capua, 28 km to the north, took the eruption as a portent of great events to come in Book 77.2.1 of his *Roman history*:

A huge fire blazed up on Vesuvius, and the roaring was mighty enough to be heard even in Capua, where I live whenever I am in Italy . . . It seemed probable that some change in the state was about to occur.

Dio Cassius had already remarked in Book 66 that the volcano had often erupted since AD 79:

The crater is given over to the fire and sends up smoke by day and a flame by night. [Many subsequent observers have made this error, because the red-hot ash could be seen only at night, as the sunlight is too bright during the day.] In fact, it gives the impression that masses of incense of all kinds are burning inside it. This goes on all the time, to a greater or lesser extent. When there is a large settling inside [and nothing blocks the exit], the crater often ejects ashes, but it throws out stones when the mountain is traversed by violent blasts of air [that were believed to be trapped inside the volcano]. It also rumbles and roars because its vents are narrow and concealed rather than all gathered together [into a single vent]. Such is Vesuvius, and these things usually happen there every year.

Unfettered by Christian theology, Dio Cassius even hazarded an opinion about the causes of the eruption. Once, he indicated, the summit of the mountain had been level, and that fire had risen from its centre, where it was constantly burning. He suggested that the constant burning of the rocks in the central core had made them so brittle that they were reduced to ashes, whereas the unburnt rocks around them had kept their original height. Because the rocks in the central core had been consumed, the whole mountain resembled an amphitheatre, from which flames and fumes were expelled.

The chief value of his description is that it shows what the mountain looked like in AD 203. The "amphitheatre" at the summit would be the caldera of Somma. Dio Cassius also noted that the summit of the mountain was level around the caldera. This was probably a misconception on his part, because, from Capua, he would see only the level northern rim of the caldera – the present Monte Somma. It is a pity that Dio Cassius apparently did not venture to the southern flanks of the volcano where the rim was lower, and where he might have noticed whether eruptions had already formed the cone of an early Vesuvius. At all events, the volcano seems to have shown some form of persistent activity until about AD 235.

The martyrdom of the Christians, AD 303–305

In AD 303, the Emperor Diocletian ordered the massacre of Christians throughout the Roman empire. His zealous agents implemented his orders over the next two years. Thus it was that the Imperial emissaries decapitated the Bishop of Benevento on the hill overlooking Pozzuoli, west of Naples, on 19 September AD 305. His saintly life and his martyrdom soon brought about his canonization as San Gennaro [Saint Januarius]. He emerged from the common lot of martyrs and became one of the most famous victims of that dire episode when the Neapolitans took his remains to their city and adopted him as their patron saint. His relics rapidly earned the reputation of being able to protect Naples in its frequent hours of need, and San Gennaro thus became one of the most renowned and revered patron saints in the world.

Soon afterwards, Constantine the Great became Roman emperor in AD 312. He transferred his capital from Rome to Byzantium on the Bosphorus, which he renamed Constantinople. As he was a Christian, Christianity duly became the official religion of the Empire and of most of its subsequent emperors. Then, when Emperor Theodosius died in AD 395, the empire was divided; and one emperor ruled in Constantinople and another in Rome. This situation came to an end in AD 476 when the Roman part of the empire collapsed, although Rome retained its role as the centre of the Latin and Catholic creed. The Byzantine part tottered along as the focus of the Greek Orthodox brand of Christianity until Constantinople fell to the Turkish armies in 1453. Both versions of the Christian faith linked volcanic activity clearly with sin. Every Christian learned the fate of Sodom and Gomorrah in Genesis 19 24–28:

> Then the Lord rained upon Sodom and upon Gomorrah brimstone and fire from the Lord out of Heaven; and he overthrew those cities, and all the plain, and all the inhabitants of the cities, and that which grew upon the ground . . . and, lo, the smoke of the country went up as the smoke of a furnace.

Every Christian, too, had ample warning about the wages of sin that were described in macabre detail in Revelations, where fire, brimstone and pits of sulphur abound. An erupting volcano could not fail to bring these texts forcibly to the mind of every Christian sinner.

The Pollena eruption, c. AD 472

Throughout most of the fifth century, the barbarians had been rampaging through Italy: the Visigoths had sacked Rome itself in AD 410 and the Vandals followed suit in AD 455. Then, on 23 August AD 476, Odoacer and his tribe invaded Italy and dethroned Romulus Augustulus, the last Roman emperor in the West. Odoacer then set up his capital at Ravenna. In retaliation, Zeno, the Byzantine emperor in Constantinople, sent against him another barbarian, the Ostrogoth leader, Theodoric the Great, who captured Ravenna and put Odoacer to death. Theodoric then himself took his chance

and founded the Ostrogoth Kingdom in Italy in AD 498. After King Theodoric died in AD 526, relations with the Byzantine emperor in Constantinople deteriorated. The Byzantine general, Belisarius, fought a successful
war against the Goths from AD 535 to AD 553. Then, in AD 568, it was the
turn of the Lombards to invade northern Italy. And so the chaos went on and
on, unabated.

Campania did not escape this turmoil. The people were forced to take refuge in their own land, repairing the damage after one horde had left and
before the next horde loomed on the horizon. What had once been "Happy
Campania" had become a land of strife, and its people had enough on their
hands without having to contend with a volcanic disaster as well. Thus, the
eruption did not occur at a happy time, and it was even more unfortunate
that it proved to be the most violent outburst of Somma–Vesuvius since
AD 79. It is also the most thoroughly studied of all its medieval eruptions, and
yet contemporary references to it are rare and its date is uncertain.

Earth scientists have called this the Pollena eruption because, near the
town of Pollena Trocchia, about 6 km to the northwest of the crater, the well
exposed volcanic deposits are 15 m thick. The outburst began with a large
column of ash, pumice, steam and gas, which quickly reached a height of
20 km. Some 0.16 km^3 of greenish pumice, and then black ash, rained down,
especially to the east, where they extended at least 30 km from the crater.

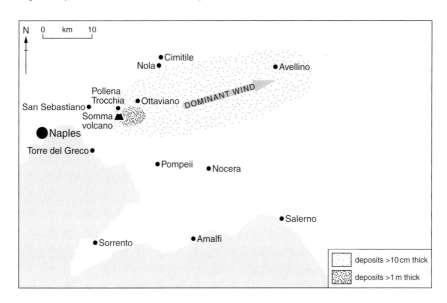

The distribution of airborne fragments from the eruption of c. AD 472.

Soon afterwards, pyroclastic flows rushed down the flanks of the volcano and spread over the plains to the north. As the eruption declined, torrential rains drenched and mobilized the loose fragments and they generated mudflows that swept widely across the northern lowlands, where they often piled up more than 1 m deep. About 0.5 km^3 of fragments was expelled altogether.

The westerly winds in the stratosphere carried the erupted cloud over the Balkans, darkening the sky and showering ash over a wide area. It reached at least as far as Constantinople, some 1200 km east of the volcano. It is this fall of ash on the Byzantine capital that has given the eruption its commonly accepted date. Count Marcellinus, chancellor to the Byzantine Emperor Justinian, noted the episode some 50 years later in his *Chronology*:

> In AD 472, when Marcianus and Faustus were consuls, Vesuvius, the burning mountain in Campania, seethed with internal fires, vomited up its consumed inner parts, brought the darkness of the night to the day-time, and covered all Europe with fine dust. In Byzantium [Constantinople], the people commemorate [the anniversary of] the arrival of this terrible dust every year on 6 November.

Most eye witnesses believed that the falling ash was a portent of deaths among high-ranking persons and disasters in significant places. Some, for instance, contended that it might be forecasting the death of the Emperor Leo I. The Byzantine emperor took a grim view of this suggestion and fled from Constantinople with his family. More courageous citizens organized religious processions to avert any dangers to the city; and they were repeated every year for more than a century on the anniversary of the event. Leo I died on 18 January AD 474, although presumably from non-volcanic causes.

This date of the ashfall seems clear enough, but other chroniclers, who make passing references to this event, place it variously between AD 469 and AD 474. On the other hand, it is not absolutely certain that the ash that fell on Constantinople came from Somma–Vesuvius, although a suitable alternative source is hard to find, because, for example, neither Etna nor Santorini, nor Monte Vulture, were apparently erupting at the time. Although gaps in the literature are by no means uncommon, the complete absence of surviving references to the eruption in Campania itself is disconcerting, since it created such widespread damage.

To add to the doubts about the exact date of the Pollena eruption, some archaeologists now believe that it could have taken place several decades after

AD 472, partly because many buildings had been damaged – apparently in the barbarian wars – before the ash fell upon them.

The eruption seems to have been unusually wet, either because it occurred in a notably humid period or because large amounts of steam were expelled. At all events, many mudflows developed, and one even reached central Naples and entombed remains that have been recently excavated at the church of San Lorenzo. It is said that the Neapolitans took refuge in the catacombs. There, weeping and wailing in desperation and anguish, they turned to San Gennaro and invoked his protection. The eruption apparently ceased at once, the ash stopped falling and daylight returned. It was enough to convince the Neapolitans that the saint had special powers and had taken a special interest in their fate. From that day onwards, San Gennaro's reputation was secure as the protector of Naples and the Neapolitans.

The eruption caused surprisingly little additional damage to buildings on the Nola plain north of the volcano. However, the eruption interrupted repair work on Roman roads and aqueducts, in the Roman amphitheatre in Nola, and in the Basilica at Cimitile near by. It also clearly stopped work on the farms, where ash and cinders covered freshly ploughed furrows and preserved them for posterity. A few dozen people, who were sheltering in a church in Cimitile, were killed when the weight of accumulated ash caused the roof to collapse on top of them. However, few other skeletons have been found in the district, which implies that the population perhaps had time to escape. The floods and mudflows seem to have lost much of their impetus once they reached Nola itself, and invaded structures such as the amphitheatre and an aqueduct near by without knocking them down. In all, this widespread absence of major volcanic destruction suggests that the Pollena eruption was not quite as violent as has been supposed, although it must have added to the hardship of the people.

However, subsequent resettlement in northeastern Campania was only sporadic and slow, although part of the amphitheatre at Nola was repaired and a new church was built at Cimitile. On the other hand, the coastal towns on the southern side of the volcano were damaged much less than the areas to the northeast, where the bulk of the volcanic fragments fell. One mudflow is known to have affected Torre Annunziata, but these settlements were populated throughout the political chaos of the sixth century, even after the Roman villas there had been abandoned.

The eruption, c. AD 507–512

If the Pollena eruption did take place a few decades after AD 472, as some archaeologists believe, it is possible that it was the same outburst that caused devastation between *c.* AD 507 and *c.* AD 512, during the reign of King Theodoric the Great. There can be no doubt that this particular eruption actually happened because the Campanians begged King Theodoric to exempt them from taxes because of the damage that it had recently caused to their homes and livelihoods. Theodoric's subjects would never have dared to invent an eruption to claim exemption from taxes levied by such a resolute monarch. And of course, they must have made their request very shortly after the eruption, when the devastation was at its height.

Their original request has not survived, but Theodoric's reply has come down to posterity in the works of the scholar who drafted it on the king's behalf. This scholar was the young Magnus Flavius Aurelius Cassiodorus, who was the king's rhetorical draftsman and legal advisor. Cassiodorus took up his appointment in *c.* AD 507 at the age of 22, but he sought other challenges before he was 30. He retired to monastic life in about AD 537, helped Saint Benedict reorganize the monasteries, and died in about AD 570.

On behalf of King Theodoric, Cassiodorus wrote to Faustus, who was the royal praetorian prefect in Campania. This letter (Letter 50 in the *Variae epistolae* of Cassiodorus) is a tantalizing document because Cassiodorus wrote in high-flown Latin, which is a challenge even to expert classicists.

Cassiodorus and Faustus held their positions together only from *c.* AD 507 to *c.* AD 512. Hence, the eruption must have taken place during or just before that period. The letter drafted by Cassiodorus offers perhaps the oldest surviving reference to the longer-term effects of an eruption on the land and people of Campania; and it is the oldest surviving indication that volcanoes give out warning signs before they erupt. In the end, Theodoric consented to reduce taxation in the devastated area, but he was not going to deplete his coffers unnecessarily. He took care to urge that Faustus, or his nominee, should make accurate assessments of all the damage before the victims were compensated. Cassiodorus expressed the royal wishes as follows:

The Campanians, devastated by the hostility of Mount Vesuvius, and having been stripped of the fruits of their fields, have poured forth tears and implored our clemency, so that they may be relieved of the burden of paying tax. Our piety rightly assents to this. But, because the misfortune of each individual is unclear to us unless it is examined, we instruct your Greatness [Faustus] to send a man of proven reliability to the territory of Nola or Naples, where poverty itself is advancing because of the damage to property. In that way, once the land there has been carefully inspected, relief may be given in accordance with the damage to the owner's livelihood. When the whole extent of the damage is recognized, let the amount of benefit be accurately matched with it.

Through Cassiodorus, Theodoric did offer his unhappy subjects a glimmer of consolation, perhaps with a certain cynicism. However, he recommended that the Campanians ought to look on the bright side as well: Campania would enjoy perfect happiness if it did not suffer repeatedly from this one cruel scourge. Moreover, he asserted, the volcano did not unleash these terrible events without due consideration: it gave out significant signs warning of what was about to happen. For some time before an eruption, "the opening of the mountain murmurs when nature is struggling with such great masses [within it], so that the wind that is stirred up terrifies the neighbourhood with resounding roaring". These calamities, then, had at least one redeeming feature: the disasters ought never to catch the people unawares, and therefore they should never lose all their possessions.

Cassiodorus provided a valuable account of the recent eruption, probably for the benefit of Faustus, who seems not to have witnessed it. His description of the emissions of seething masses and rivers of dust indicates that the volcano had erupted pyroclastic flows, possibly associated with mudflows:

The air there is darkened by the most foul exhalation, and, when that destruction is set in motion, it is recognized virtually throughout the whole of Italy. Burnt-out dust flies through the great void and, when clouds of earth have been stirred up, the dusty drops rain down on the provinces overseas as well . . . You would see [in Campania], as it were, some rivers of dust in motion, and, as it were, liquid currents rush down in a sterile seething mass [probably pyroclastic flows?]. You would be amazed that the surface of the fields had suddenly swollen up as high as the tree tops; and what had been coloured the most fertile green was suddenly devastated by the calamitous heat.

Cassiodorus had no better idea about how a volcano really worked than any of his contemporaries, but his comments reveal a glimpse of the nature of the eruption that had just taken place:

What is this extraordinary anomaly whereby one mountain should rage
in such a way that it is proved to terrify so many sections of the world
. . . and to scatter its substance everywhere, without seeming to suffer
any loss [of volume]? It showers down dust far and wide and belches out
great lumps on the local people, and has been regarded as a mountain
for so many generations, although it is consumed by such great emissions. Who would believe that such huge chunks have burst out and
have been carried from such deep caverns right down to the plains; and
that, as it were, light chaff has been thrown out, spewed from the mountain's mouth by an exhaling wind? . . . Virtually the whole universe has
been granted knowledge of the fires of this mountain. How then could
we not believe the inhabitants? . . . Therefore, as has been said, let your
[Faustus's] prudence choose such a man as can both bring relief to those
who have suffered harm, and leave no room for fraud.

Like Strabo before him, Cassiodorus did realize that many areas covered
by volcanic ash seemed to recover their fertility within a couple of years of
appearing to be totally devastated. This valid point also offers an additional
indication that the people's request for tax exemption clearly must have been
made very soon after the eruption:

That perpetual furnace spews out sand formed from pumice, but it is
still fertile. Although the sand [ash] has been dried up by lengthy burning, it soon puts forth shoots that develop into various plants, so that
what had been devastated shortly before is restored with great speed.

King Theodoric duly exempted the victims from taxes for several years, by
which time the devastated land would probably have recovered much of its
original fertility.

The historian, Procopius of Caesarea, in Palestine, was most probably
referring to this violent eruption when he commented upon the behaviour
of the volcano in his *History of the wars: the Gothic War*. Procopius was the
secretary to Belisarius, the Byzantine general who was leading the campaign
against the Ostrogoths in Italy in AD 536–540.

Procopius described the effects of ashfalls and the growth of a tall column
of ash and fumes. When the volcano expelled great masses of ash, Procopius
recommended that it would be better not to approach too closely.

If, by some mischance, any passer-by were to be taken by surprise by this terrible shower, he could not possibly survive. And if the ash were to fall upon houses, they too would collapse under the great weight of the mass. Whenever a strong wind blows up, the ash rises to a great height, farther than the eye can see, and it is transported wherever the wind might drive it. It then falls down on lands a great distance away . . . It is said that this rumbling used to occur every hundred years or more, but latterly it has happened more often.

A possible eruption in AD 536?

In Book 6.4.21–30 of *The Gothic War*, Procopius of Caesarea developed the theme, recently mentioned by Cassiodorus, that volcanoes gave out preliminary signs that they were about to erupt. He noted that the volcano did indeed give out warning signs in *c.* AD 536. Paradoxically, the preliminary rumbling proved to be a rather unusual false alarm. The volcano did not erupt:

> The mountain of Vesuvius rumbled, and although it did not break out in an eruption, the rumbling still led the people to expect with great certainty that there would be an eruption. And for this reason, it happened that the inhabitants [living near by] fell into a great terror . . . When the mountain gives out a rumbling sound like roaring, it generally sends up a great quantity of ash not long afterwards.

Although the volcano missed the chance to shine in the presence of an intelligent observer, Procopius was able to provide a valuable, but regrettably incomplete, description of Somma–Vesuvius in about AD 536–540. The crater, he wrote, was very deep, and molten lava was still playing in the abyss, which had no doubt been formed by the previous violent eruption, and this had led the people to claim tax exemptions. However, he makes no mention of any cone within the great central crater of the mountain:

> This mountain is 70 stadia from Naples . . . It is a very steep mountain, whose lower parts are spread widely around it and are covered by delightful vegetation, whereas its upper part is precipitous and exceedingly hard to climb. At about the centre of the summit, a chasm can be seen, which is so deep that it seems to extend all the way down to the roots of the mountain. Anyone daring to peer over the edge of the chasm and look down into its terrifying depths would see the fire. As

How many eruptions occurred between *c.* AD 472 and *c.* AD 536?

- At least one major eruption took place. The evidence for this is the letter drafted by Cassiodorus on behalf of King Theodoric. The role of Cassiodorus means that this eruption must have taken place between *c.* AD 507 and *c.* AD 512, or very shortly before. No deposits have apparently been attributed to this eruption. Two short references in the Pascale Companum also indicate that the volcano erupted on 9 November AD 505 and that it had darkened the sky near the mountain on 18 July in AD 511.
- The Pollena eruption certainly took place. The evidence is the thick volcanic deposits which are often found overlying archaeological remains around the northern fringes of Somma–Vesuvius. The eruption has been dated, rather indirectly, from a fall of ash on Constantinople in *c.* AD 472. Some archaeologists believe that this eruption could have taken place several decades later. The Pollena eruption and the *c.* AD 507–512 eruption could therefore be one and the same.
- Procopius says that the volcano rumbled in *c.* AD 536, but he states explicitly that it did not erupt. His descriptions of volcanic activity could probably be referring to the eruption described by Cassiodorus.
- The most reasonable conclusion would therefore seem to be that the chief – and perhaps the only – eruption of Somma–Vesuvius in the period occurred between *c.* AD 472 and during, or shortly before, the interval *c.* AD 507–512. This rather violent eruption expelled volumes of ash and possibly pyroclastic flows that devastated the countryside, and might even also have emitted a lava flow.
- If there were indeed two eruptions, then one must have taken place about *c.* AD 472 or a decade or two later, and the second about *c.* AD 507–512.

a rule, however, the flames merely twist and turn upon one another, and they thus cause no trouble to the inhabitants in the region.

In Book 8.35.1–6, Procopius turned his attention to a smaller eruption on Vesuvius. It is possible here that he was presenting the oldest surviving account of the emission of a lava flow from the volcano, but it seems equally possible that he was describing a flow on the flanks of Etna. Nevertheless, he did add that the flows were exactly the same as those on Somma–Vesuvius. Hence, some lava flows must have been emitted from Somma–Vesuvius before *c.* AD 536–540, although they were not recorded in any surviving text:

Fairly large and small stones are thrown from the entrails of Vesuvius and spread all around its base. And at the same time, a torrent of fire runs down from the summit to the base of the mountain and beyond . . . In breaking [from?] the ground, the torrent of fire forms high banks on either side, cutting into the ground. As the flame is taken along in the channel, it looks like a flow of burning water at first. However, as soon as the flame is extinguished, the stream stops running at once and goes no farther. And what remains of the fire looks like ashy mud.

The presence of a lava flow is indicated by the high banks that commonly form when the lava cools and solidifies along its edges, although molten lava continues to run along its axis. However, Procopius then confuses two sorts of volcanic emissions, which implies that he had not witnessed the events he mentioned. A lava flow solidifies with a billowing or jagged surface, like clinker or cast iron, but it looks nothing like "ashy mud". The most likely cause of "ashy mud" would be either a mudflow or a pyroclastic flow. The context thus suggests that he was making a general statement about the behaviour of the volcano from known sources rather than describing a specific event. Unfortunately, Procopius did not give the date of this activity, but his accounts do not add up to a firm statement of an eruption in *c.* AD 536. On balance, his comments would seem to be best attributed to the eruption previously described by Cassiodorus.

The eruption in AD 685

For most of the sixth and seventh centuries, the volcano was probably either inactive or was erupting too weakly to attract the attention of local scholars. Somma–Vesuvius sprang back to life when Benedict II was pope in AD 684–685. It was held that the eruption was foretelling the deaths of the pope and other famous persons (who, not unnaturally, often died soon afterwards anyway). The event received a short notice in Paul the Deacon's *History of the Lombards*, written in the monastery of Monte Cassino about a century after the event. In addition, a Greek account of the life of San Gennaro mentioned that strong earthquakes shook Campania during the winter of AD 684–685, which could have been a sign that magma was once more rising beneath the mountain. At the end of February AD 685, the volcano gave off large clouds of ash that were riddled with flashes of lightning; ash and dust rained down and burned the fields all around the mountain and reached the outskirts of Torre Annunziata. "Glowing and turbulent rivers of fire" spread as far as the sea. They were most probably small pyroclastic flows, although lava flows could not be entirely excluded, because, for example, the snout of a lava flow lying beneath the town hall in Torre del Greco has sometimes been attributed to this eruption. At all events, the eruption did not last long; it reached its brief climax in March AD 685. Pope Benedict II died on 8 May AD 685.

More persistent activity, c. AD 787–1139

In about AD 787, Somma–Vesuvius began a new regime that was to last until AD 1139. The magma that rose into the volcano was more fluid and hotter than it had been for many centuries, and thus the gases contained within it could escape more freely and with fewer explosions when they reached the crater. The main vent probably stayed open for much of the time, so that less violent and more effusive and voluminous emissions of molten lava were more important than for many centuries. Indeed, more eruptions probably took place than have yet been brought to light. Lava fountains and clots of spatter, ash and cinders gushed skywards and fell back to the ground to accumulate in a cone around the main vent, and copious lava flows invaded the old caldera and spilled over its lower southern rim, streamed down the southern flanks of the mountain, and sometimes entered the sea. Moreover, several important eruptions also burst out from vents that were low on the southern flanks of the volcano, which placed the coastal settlements in added danger. On the whole, the eruptions were more constructive than their violent and destructive predecessors.

These effusions expelled many of the older lava flows that reached the coast between Portici and Torre Annunziata. Archaeomagnetic studies have established the dates of these emissions, and it is therefore rather ironic that so few surviving texts ever even refer to them. It is particularly surprising that religious scholars should not have given such apparently doom-laden events greater prominence as the millennium approached.

The eruption in AD 787

After resting for a century, Somma–Vesuvius indulged in a brief spell of renewed activity that lasted from October to December in AD 787. It was an explosive eruption that piled up ash and clinker around the vent, but was also one where lava flowed out from the crater and down towards the sea. Thus, if the cone of Vesuvius had not already started to form within the caldera in AD 685, it would almost certainly have developed in AD 787.

Eruptions in the late ninth and early tenth centuries

Recent archaeomagnetic studies have revealed a flurry of eruptions that despatched lava flows down towards the Bay of Naples between c. AD 850 and c. AD 950, which previously had scarcely been suspected because they had not been mentioned in surviving contemporary texts. One of the earliest of

Archaeomagnetism

When Earth scientists began to study the history of volcanoes, they found that the different eruptive episodes were often very hard to disentangle by conventional geological methods. For instance, volcanoes are commonly more difficult to analyze than, say, periods of calm marine sedimentation, where successive beds, often containing datable fossils, are laid down in succession, one on top of the other. Volcanic eruptions, on the other hand, produce rocks that rarely contain any fossils, and they can be laid down side by side or in a chaotic succession that is often difficult to interpret. The problem is made more acute because, at different times, a volcano may often expel rocks with very similar chemical and physical properties.

Archaeomagnetic studies have been developed as a way of dating rocks by means of the magnetism that becomes fixed in hot volcanic rocks when they are laid down. The basis of these studies is that, although the geographical North Pole of the Earth does not change, the position of the magnetic North Pole is constantly changing. At present, for instance, a freely swinging compass needle points to the magnetic North Pole, which lies in northern Canada, 1900 km from the geographical North Pole. The angle that the needle makes with the direction of the geographical North Pole is the declination of the magnetic North Pole. The compass needle also dips downwards at an angle, or inclination, determined by the latitude of the place where it is situated. When hot volcanic rocks cool and solidify, they acquire the magnetization that is determined by the Earth's magnetic field at that moment. Once it is acquired, this magnetization does not change thereafter. It can now be revealed by very careful laboratory analysis of well chosen uncontaminated specimens, whose exact orientation in the field has been measured with extreme care.

Once the volcanic samples have been analyzed, the parameters of this magnetization from different rock samples of different ages can then be plotted onto a graph, where the vertical axis is the mean inclination, and the horizontal axis is the mean declination (see overleaf). Thus, the plotted positions of rocks that solidified at the same time on the same volcano will occur at the same point on the graph. The plotted positions of rocks that solidified at different times will, of course, occur at other points on the graph. Once many good samples have been plotted, the positions of different rocks of different ages will form a curve on the graph that illustrates the various positions of the magnetic North Pole when these volcanic rocks were laid down. Therefore, the curve also shows the relative ages of these rocks. Additional specimens of volcanic rocks can then be dated quite accurately by plotting their magnetic data onto this archaeomagnetic curve.

When an archaeomagnetic curve has been established for a volcano, it becomes an invaluable tool for dating the episodes in its history. These archaeomagnetic studies bring order to the volcanic succession, and offer scientific substance to the chroniclers' allusions to volcanic activity during the Middle Ages.

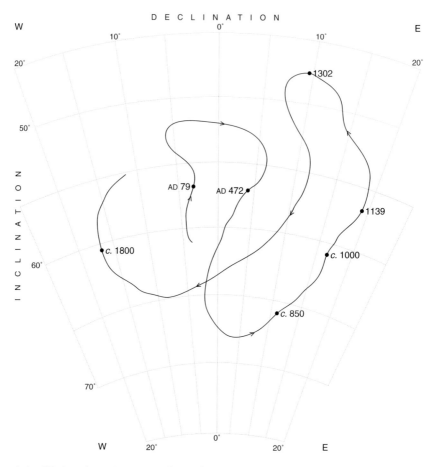

A simplified secular-variation curve from archaeomagnetic studies of lavas erupted from Somma–Vesuvius.

these emissions erupted in the Cognoli Valley, high on the southeastern flanks of the volcano, whereas the lavas from another eruption formed Le Mortelle point east of Torre del Greco. Perhaps a few decades later, other lava flows were emitted that now form some of the small cliffs and promontories between Portici and Torre Annunziata. A vent low on the southwestern flanks of the volcano, for example, built the twin Tironi cinder cones (which have since been removed by quarrying) and gave off the lavas that reached the coast at Granatello and at the Punta Quattro Venti near by. A similar eruption low on the southeastern flanks of the volcano formed the cinder cone at the Bosco del Monaco, 1.5 km from Torre Annunziata, and sent out a broad lava flow that reached the coast just west of the town.

The eruption in AD 968

The eruption in AD 968 earned the briefest of mentions when an anonymous monk, writing in Monte Cassino, observed that San Gennaro intervened to save Naples during an eruption in about AD 960. However, there is an altogether more picturesque reference to the eruption in AD 968 in an account that was written about a century later by Saint Peter Damien (1007–1072). This learned prince of the Church was a cardinal, Bishop of Ostia, a founder of hermitages, a reformer, a fervent exponent of doctrinal purity and orthodoxy, a scourge of sinners (especially among the rich and powerful), and a very medieval believer in the imagery of portents.

The source of Peter Damien's account is a holy man, who lived in a cave near Naples. One day the man noticed, on the road below, a group of black men transporting a large load of hay. They looked like Ethiopians, which, in accordance with the beliefs of the day, made him suspect them as servants of the Devil. "Why are you carrying all that horse fodder?" he asked. "We are evil spirits," they replied, "and our straw is not to feed animals, but to make a fire to burn some men – Pandolfo, the Prince of Capua, who is dead already, and Giovanni III, the Duke of Naples . . . who is still alive".

It seems that the holy man kept aristocratic company, for it turned out that he found himself soon afterwards in the presence of Duke Giovanni, to whom he told his tale. Giovanni saw the portent at once and declared that he would don the monastic habit as soon as he had explained the situation to Emperor Otto II, who was on his way southwards to fight the Saracens. Nevertheless, to be on the safe side, Giovanni took the precaution of sending an agent to Capua to verify what the Holy man had said. When the agent reached Capua, he discovered, of course, that Prince Pandolfo had already died. Giovanni himself survived for scarcely two weeks and expired before the emperor reached the area.

Whereupon, Vesuvius erupted, seemingly using the hay that the evil spirits had brought. As always, the reprobates had got their just desserts. So much "resinous and sulphurous material emerged" that it formed a flow that rushed towards the sea. To the holy man, this lava flow seemed perfectly to describe the fire and sulphur in which sinners lay in Revelations 20.14.

Saint Peter Damien was also eager to cite the dire fate of other powerful contemporary sinners. He recounted that, as soon as the eruption had started, the Prince of Salerno, had declared: "No doubt some scoundrel is going to die soon and go to hell". And lo, the following night, the Prince of Salerno himself expired, clasped in the arms of a prostitute. Even those in holy orders

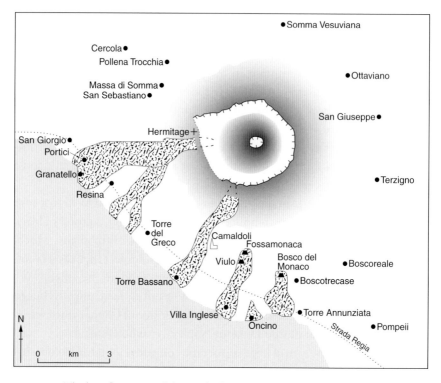

The lava flows emitted during the late ninth and early tenth centuries.

were not immune. A local priest had been keen to learn more about where the volcanic materials were coming from. He celebrated Mass, climbed up the volcano, and ventured farther than even a brave man should have done. But he could not bring himself to turn back, and he was never seen again. The volcano was all too clearly an entrance to hell.

The volcanological information that can be gleaned from these tales suggests that lava flows constituted the most striking aspect of this eruption. The mention of "resinous and sulphurous materials" apparently marks the first surviving unequivocal reference to lava flows, and they seem to have been unusually fluid because they reached the sea at some speed. Although ashfalls and pyroclastic flows might also have formed part of the "conflagration", they probably played a minor role in this eruption. Archaeomagnetic evidence suggests that lava flows from this time now form the plinth of the Torre Bassano on the coast, due south of the volcano.

The eruption in AD 999

Saint Peter Damien also provided the chief textual source for an eruption of ash, cinders and lava flows that took place in AD 999. Archaeomagnetic studies suggest that this eruption probably formed the cones of Viulo and Fossamonaca, which lie on the lower flanks of the volcano about 3 km northwest of Torre Annunziata. One lava flow breached the seaward sector of the Fossamonaca cone and reached the shore west of Torre Annunziata, where it is now exposed as the lower layer of lava in the Villa Inglese quarry and at the promontory forming the base of the Torre Scassata near by.

The eruption in 1007

The textual evidence for the dates of the eruption in 1007 is slight; it may have happened seven years before or after the millennium. It is based on a brief reference made in his *History* in 1030–1047 by Radulphus Glaber (Ralph the Beardless), a monk based in the Abbey of Cluny, in France, who had never visited Campania. He described how enormous rocks were thrown out with such power that they fell back to Earth no less than 5 km away. Obnoxious gases, mixed with fires produced from sulphur, were also expelled and they gave off such a fetid stench, probably from hydrogen sulphide, that the surrounding area was made uninhabitable. The archaeomagnetic studies suggest that the lavas forming the base of the Oncino tower at the Scogli di Prota promontory, near Torre Annunziata, could have erupted at this time.

The eruption in 1037

The evidence for the eruption of 1037 stands on firmer foundations than most of those that had taken place since AD 79. The chief reason for this might be because it occurred low down on the flanks of the volcano, only about 4 km northeast of Torre del Greco, and was thus easier to see than eruptions higher up the mountain. Activity seems to have started on 27 January and it continued well into the month of February. The eruption sprang from three vents that were aligned along a fissure running from east to west. Ash and cinders built the three Monticelli cones and, at the same time, streams of lava made their way down to the southern coast, where they formed the upper lava flow exposed in the Villa Inglese quarry, and damaged the village of Callastro, near Granatello. Writing in 1632, the Spanish commentator Simon de Ayala added a kind of confirmation of these events when he said that this eruption had formed "a great river of flames" composed of "liquid fire", which surged from the mountain and entered the sea.

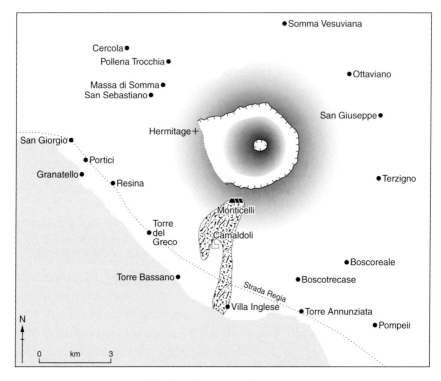

The lava flows emitted in 1037.

An eruption in 1068–1078?

The exact date, and the very existence, of the eruption in 1068–1078 are both in doubt. The apparently reliable historian, Amato of Montecassino, mentioned an eruption of the volcano in his *History of the Normans*. The text implies that the eruption must have taken place between 1068 and 1078. Amato described how a great mouth opened up on Vesuvius:

> Such was the quantity of ash expelled that it covered all the surrounding province, almost all of Calabria, and even parts of the land of Apulia. On the slope of the mountain a cavity appeared that no-one had ever seen before, and a river of boiling water poured from it continuously for 15 days. Wherever this water flowed, it dried up and burned the ground and the trees with its heat.

The ash and the crater cavity were common occurrences, but the "river of boiling water" presents a problem. They just don't make rivers like that any more. The explanation probably lies in the fact that medieval scholars could not conceive that anything in nature could flow, apart from water and molten metals. As a result, "the river of boiling water", which was hot and burned trees, seems most likely to have been a lava flow that scorched the surrounding vegetation.

As for the exact date of these events, Amato's description could fit the eruption in 1037, but, since he was writing in 1078, an error of almost 40 years during his own lifetime would seem rather excessive for an historian. All in all, then, it might be best to give Amato the benefit of the doubt and allow him his separate eruption sometime between 1068 and 1078.

The eruption in 1139

The eruption in 1139 was dated with unusual precision. It began on 1 June with violent explosions that gradually decreased in power during the next four weeks until activity ceased altogether on "the feast of Saint Peter and Saint Paul" (29 June). Surprisingly enough, two commentators agree not

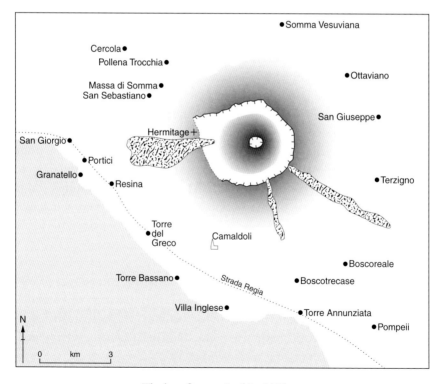

The lava flows emitted in 1139.

only on the dates but also on what happened: one was an anonymous monk writing in the Abbey of Cava dei Tirreni, near Salerno; the other was Falcone Beneventano, who was then a judge in Benevento, northeast of the volcano, and who later became secretary to Pope Innocent II. Beneventano reported in his *Chronology*:

> In this year [1139], on the first day of June, the mountain that can be seen near Naples threw out powerful fire and bright flames for eight days, with the result that all the city and the neighbouring villages were awaiting death. A black and horrible dust came from this conflagration and spread as far as Salerno, Benevento, Capua and Naples . . . and [it piled up so thickly that] the people of Benevento did not see the [former] ground [surface] for 30 days.

Archaeomagnetic studies have identified three main lava flows from this episode, which seem to have come from the main crater. On the western flanks of the volcano, one tongue forms a wide lobe above Portici and Granatello, which solidified when its snout reached a height of only 170 m above sea level. Another flow spread down the southeastern flanks of the volcano into the area between Boscotrecase and Terzigno; the third and shorter flow extended into the Cognoletto Valley.

By the end of the eruption in 1139, almost 400 years of repeated mild explosions of ash and cinders, and prolific effusions of lava, had choked the floor of the old caldera of Somma and built up the cone of Vesuvius within it. Vesuvius had become a distinct, and clearly separate, volcano rising from the plinth of Somma; and it was to remain the focus of attention and of most of the eruptions until the present day. In 1139, the physical features, if not of the human occupation of the volcano, must have looked very much like the view from Naples today. Vesuvius then took a rest. For a time, the focus of volcanic attention shifted westwards.

The eruption of La Solfatara in 1198

La Solfatara rises in the Campi Flegrei on the eastern outskirts of Pozzuoli. In about 1800 BC, explosions of fine ash had blasted out a deep broad crater, which became a centre of continued mild activity that reached a very modest climax in 1198.

Perhaps the unusually cool and humid climate in the region during the preceding centuries had raised the height of the water table under the crater.

Then, one of the many earthquakes in Campania could have disturbed, or even blocked, the circulation of the groundwater beneath the crater until pressure built up enough to release it forcibly once again. In 1198, the hot water, steam, gases and decomposed rocks burst to the surface in a muddy geyser that spurted 10 m or more into the air. The geyser may have gushed for several hours or repeated its display at intervals during the succeeding days or weeks. Then La Solfatara resumed its more usual tranquil activity, with just enough bubbling and hissing to conjure up yet another vision of Hell in the susceptible medieval mind.

The eruption of Arso in Ischia in 1302

After the small eruptions on Ischia that took place during the latter years of the Roman republic, the island stayed quiet for more than six centuries. Suddenly, sometime between AD 670 and AD 890, a powerful explosion took place near the present village of Fiaiano, about 2 km south of Ischia Porto. Although it covered the northeastern part of Ischia with a blanket of pumice, the whole event probably lasted less than a couple of days before the island calmed down again for another 400 years.

In 1275, an earthquake rocked Ischia and might have stimulated an

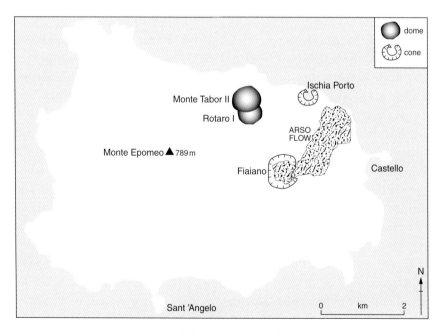

Ischia and the Arso eruption of 1302

upward movement of magma. In 1302, the very same volcanic vent that had expelled the Fiaiano pumice provided a pathway that enabled another batch of magma to reach the surface. "When Pope Boniface was alive . . . in the reign of Charles II of Anjou . . . Ischia threw out a great fire from its sulphurous cavern".

The eruption began on Thursday 18 January 1302 and built what later became known as the Arso volcano ("the burnt place"). A vigorous explosion expelled ash and pumice that devastated the crops throughout the most fertile eastern parts of Ischia, rained down onto the nearby mainland around Cumae, and choked the shore. A vast cloud soon extended all over Campania, and showered ash up to 75 km away, as far east as the town of Avellino and the Abbey of Cava dei Tirreni near Salerno. The whole region was covered with an ashen blanket that looked like snow. The Dominican monk, Bartolomeo Fiadoni, described how:

> There was a great cloudiness over the region and the sea was filled with burnt stones and with burnt terrestrial materials called pumice, which float on the sea because of their porosity [which means that they can] contain air . . . Ashes were formed in such enormous quantity that they seemed like mountains in their bulk.

Activity changed character even before the first day was out. "Ischia began to burn and the people took flight and left". They seized every available boat and made as fast as they could for the neighbouring islands of Pròcida and Capri, and eventually for Naples, Baia and Pozzuoli on the mainland.

Lava splattered from the vent and piled up in the cone of Arso, which soon reached about 50 m high and 450 m wide. At the same time, lava also gushed out in a wide stream resembling molten metal that entered the valley of the Rio Corbore and flowed slowly but unstoppably northeastwards. It reached the sea, hissing and crackling like tempered metal in the cool waters, and formed the Punta Molina, a small fan-shape promontory 1 km wide, just east of Ischia Porto. The climax of the outburst was quickly spent, although it spluttered on for another two months.

In 1471, Giovanni Pontano (1426–1503), secretary to King Ferdinand I of Aragon, discovered some eye-witness descriptions of this flow when he was doing research in the Neapolitan archives of that period. As he wrote later, the flow from Arso had been just like those that Virgil and Pindar had described on Etna in antiquity:

In fact, stones of a large size lie both near the sea and in the fields along-side this flow. In places, great rocky reefs emerge on the coast and in the sea, which are so burnt and worn that they look like liquid foam even today. There are also flows of stones like those described by Pindar, which look like foam rather than the liquid matter that usually resembles molten bronze . . .

The islanders had succumbed to panic long before the snout of the lava flow entered the sea. Northeastern Ischia was devastated for several decades, and for centuries afterwards the barren lava flow itself was known as *le cremate* (the incinerated areas). Some claimed that many people and animals had met their deaths in the disaster, although the number of human victims is unlikely to have been high, because this was a rather modest eruption that happened in a rural area. The people who survived the initial blast of pumice would easily have been able to run away from the advancing lava flow. Although this eruption could well have terrified the Ischians, it was far from measuring up to any eruption of Vesuvius. But Vesuvius, of course, was resting.

Dormant Vesuvius

For more than a century after Vesuvius had erupted magma in 1139, it no doubt behaved like many other dormant volcanoes. The main volcanic vent would have been choked by solidified lava so that only small quantities of gas and fumes could rise into the crater from the mass of magma then lying perhaps 4 km or 5 km below the surface. In their day, both Saint Thomas Aquinas (1225–1274) and Petrarch (1304–1374) mentioned some volcanic emissions, but hard evidence is very scanty. For instance, Petrarch's reference to an eruption might just possibly be related to general tales of gloom related to the spread of the Black Death (1347–1350), which exterminated far more people in Europe alone than all its volcanoes had killed for thousands of years. No less an expert than Giovanni Boccaccio offered a contrary testimony. From about 1355 to 1374, the author of the *Decameron* had turned his attention from human to environmental behaviour. He affirmed that, "after having erupted burnt stones for so many centuries, nowadays neither flames nor fumes escape from its crest, [but] there is a great opening [at the summit] that clearly bears witness to the conflagrations of the past". He also noted that the mountain was already clothed everywhere in vineyards and orchards, which indicated that many decades must have passed since the most recent eruptions had destroyed the farmlands on its lower slopes.

Nevertheless, Vesuvius may have coughed in 1500. Sometimes, groundwater seeps down into a dormant volcano and reaches the magma rock lying at depth and generates a vigorous hydrovolcanic eruption. Such explosions can last no more than a few minutes, or for a few days at most. Such an explosion may have happened on Vesuvius in 1500. In his history of Nola, Ambrogio Leone wrote that:

> For three days, the air went dark to such an extent that the stupefied people began to tremble. When the eruptive violence that had thrown out the volcanic materials had come to an end and abundant reddish ash had rained down, everything was covered as if with fine snow.

Although the area around Nola was quite densely populated, no other account of this eruption has apparently survived, or perhaps has ever existed. It seems that the author had been absent in Venice at the time and had gained this information – or misinformation? – by hearsay. His description could fit many other medieval eruptions, and he could simply have copied it from older sources, or he may merely have been repeating old folk tales.

A second view of events on the volcano at the time came from the Spaniard, Gonzalo Hernandez Oviedo, who later went to the Spanish Americas and made his name as the author of the *Historia general de las Indias*. In 1501, he spent three days on Vesuvius, but the volcano rewarded him with nothing more than a vast quantity of fumes. However, he did comment that some people claimed that these fumes glowed brightly during the night. If this was true, then it is possible that magma could have been lying close to the surface and might have caused an eruption the previous year. Indeed, the copious fumes themselves suggest that the volcano was more active in 1500–1501 than it might have been since 1139. At all events, the volcano soon resumed its torpor. As the sixteenth century drew to a close, Vesuvius had been dormant for over 450 years. It was time for it to show its mettle again.

Further reading

Dio Cassius 1961; Gasparini & Musella 1991; Principe et al. 2004; St Peter Damien 1050; Tanguy et al. 1997, 1999, 2003.

Chapter 6

The eruption of Monte Nuovo:
a new approach

While Vesuvius lay dormant, the volcanic centre of action shifted briefly to the Campi Flegrei. Eye witnesses described and tried to explain the eruption with the new curiosity inspired by the Renaissance.

Vesuvius had fallen dormant and La Solfatara had continued to bubble, but the people of the Campi Flegrei soon had another, and much more serious, cause for alarm. During the late fifteenth and early sixteenth centuries, the very land itself became increasingly restless: the ground in and around Pozzuoli began to rise. On 29 September 1538 magma surged from the ground.

Entirely new mountains do not form very often within a human lifespan, and still more rarely do they appear almost overnight. Those who witnessed the event in 1538 did not even have the benefit of previous records to help them understand what had befallen them. They were puzzled, astonished and afraid. Fright seems to have dulled their imagination to such an extent that they named the new mountain . . . Monte Nuovo. It was a modest eruption, with nothing like the power displayed in virtually every outburst of Somma–Vesuvius. Yet the birth of Monte Nuovo was the most powerful natural event that had taken place in Campania for 400 years. It shook the town of Pozzuoli severely and buried a village under a cone 134 m high.

An intellectual change

The accounts of the eruption of Monte Nuovo are very different from those that have survived from the eruptions from Somma–Vesuvius during Medieval times. The eye witnesses reveal a distinct change of attitude to the features of the natural world that was in keeping with the humanism and new

A woodcut of the eruption of Monte Nuovo in 1538, seen from the sea, from the account published by Marco Antonio Delli Falconi later that year. The features named are, from left to right: Miseno, Baia castle, the sweating-rooms of the so-called Baths of Nero, Monte Nuovo, the shoreline before the eruption, and the ancient molo (jetty) of Caligula, Monte Barbaro, Pozzuoli and La Solfatara.

philosophies of enquiry that marked the Renaissance. Their culture was no longer exclusively biblical, but classical and Italian. In that sense, they offer the first fairly reliable descriptions of an eruption in modern times.

The authors of these accounts were people who had been educated, at least in part, in the new ways. Of those whose profession is known, some had trained as doctors, such as Toleto and Porzio; others, such as Delli Falconi, were typical Renaissance priests, who had studied natural sciences and were steeped in the literature of antiquity; and Miccio was the Spanish viceroy's biographer. The profession of other commentators, such as Del Nero, Russo and Marchesino, is unknown, although Marchesino must have had a religious background.

They all wanted to record the eruption as a feature that could, and should, be explained in terms of the natural workings of the Earth. They tried to describe what had happened; they estimated sizes; they described the shape of the new landforms; they assessed the extent and the nature of the devastation; and they attempted to count the casualties. Moreover, the impact of

116

the eruption on the people was not seen merely as an incitement to prayer and the confession of sins, but as an event with a human impact, causing panic and destroying homes, crops and livelihoods. Moreover, in order to make their accounts more graphic, they sought to make comparisons with the more familiar world around them. Thus, the noises of the eruption were sometimes compared to thunder, but more often to the loudest noises that they could conceive – the sounds of war: artillery, gunfire, and the most violent of battles. Some eye witnesses even went so far as to try and discover what natural features might have caused the eruption and how they operated. Nevertheless, the commentators still suffered from certain handicaps. It is very unlikely that any of them had ever set eyes on an eruption before. Like their medieval predecessors, they had no technical or scientific vocabulary, and they still did not even have the word "volcano" that would have enabled them to focus their attention on the specific nature of the features they saw.

Spanish rule

In 1538, the Spaniards were now in charge of Campania, Sicily, and most of southern Italy. Charles V was King of Spain and also Holy Roman emperor. The task of administering southern Italy fell to a viceroy, nominated from the grandest of the grandees of Spain. One of the ablest of all these viceroys was Pedro de Toledo, the marquis of Villafranca, who ruled from 1532 until 1553. It was an important post, for at that time Naples, Paris and London were the three most populous cities in Europe, each with some 250 000 inhabitants. Pedro de Toledo took full advantage of his unusually long reign to transform the urban fabric of central Naples and make it one of the finest capitals of the period. Indeed, he gained such a reputation that his daughter, Eleanora, married Cosimo I the Great, the Duke of Florence.

Pozzuoli and Tripergole

Pozzuoli, out along the northern shores of the Bay of Naples, was a town of about 22 000 inhabitants. It had been the main centre in the Campi Flegrei for over 1500 years, although its vast Roman amphitheatre, which had been the third largest in the empire, and its old Roman market had long ago fallen into ruin. On the other hand, the port was flourishing, with the harbour

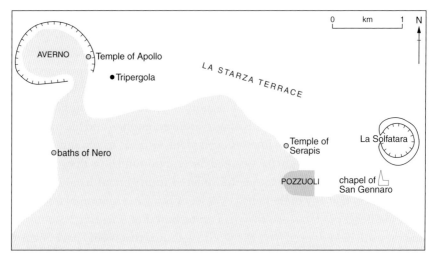

The north shore of the bay of Pozzuoli before the eruption of Monte Nuovo, with the crater of Averno volcano forming a rounded inlet on the coast.

snuggled under the headland that carried the cathedral, the churches and the main municipal buildings of the town. It was a pleasant place to live and Pozzuoli enjoyed the same locational advantages that had graced Roman Herculaneum – but farther from the dangerous volcano, or so it seemed.

The Campi Flegrei, near by, formed a fertile lowland, scattered with small villages and dotted with low hills, old volcanic craters and a few lakes. About 5 km west of Pozzuoli stood the small spa town of Tripergole, dominated by a thirteenth-century castle crowning the Monticello del Pericolo ("hill of danger"). Tripergole had been built around hot springs issuing from some of the many faults that fissured the whole region. Their curative properties had been exploited for centuries, and a hospital with 30 beds had been set up in 1307. Like many others spas, Tripergole soon attracted a pharmacy, a clinic, baths and sweating rooms. At times, even foreign visitors formed part of the throng waiting for treatment. As Antonio Russo of Pozzuoli later reported, three inns catered not only for the infirm, but also for "the gentry and rich people with money to spend". At Whitsun, large happy groups of ordinary people from Pozzuoli used to walk out along the old Via Domiziana to the fiesta at the church of Santo Spirito in Tripergole, where they all danced and ate the cherries and bread rolls provided by the authorities.

The fall and rise of the Temple of Serapis

The ruins commonly known as the Temple of Serapis at Pozzuoli illustrate perhaps more clearly than any other building in the world how the relative levels of land and sea have changed since Roman times. They have been famous in the scientific world ever since an illustration of them formed the frontispiece for Charles Lyell's *Principles of geology* (1830), which ran to many editions. The function of the original building had been misinterpreted when a stone bearing an inscription to the Egyptian god, Serapis, was found among its ruins. It is, in fact, the remains of a Roman market built in the first century. Soon after this market was completed, the level of the land sank down to such an extent that the sea flooded into the building. Commercial necessities ensured that the market had to be resurfaced with a new pavement, some 2 m higher than the original. Eventually the market was abandoned during the chaos that accompanied the fall of the Roman empire in the fifth century. Much of the building collapsed, and rubble covered the lower 3.6 m of the columns that had stayed upright. However, the land went on sinking, so that by the tenth century, the building was some 5.8 m lower, and the sea had flooded the lower parts of the columns that protruded above the rubble. These submerged parts of the columns were then attacked by a mollusc called *Lithodomus lithofagus*, which bores into stone and gives them a characteristic deeply pitted surface. The upper parts of the columns were never submerged and were therefore exposed to the air.

Then, little by little, the land began to rise again during the fifteenth century. After a while, the floor of the market rose above the waters, and revealed the once-submerged surface of the columns that had been pitted by the mollusc. The clearance of rubble revealed the lowest parts of the columns that had been protected from the attack of the mollusc, thereby remaining in an almost pristine state. The columns became a kind of register of the fluctuations of the land: their upper parts had been slightly weathered during centuries of exposure to the air, the middle reaches had been pitted by the mollusc when they had been submerged, and the base had been well protected by the rubble on the floor of the market.

The changes in the relative level of the land and the sea took place in response to fluctuations in the volume and depth of magma lying below the surface of the region. These are called bradyseismic movements (from the Ancient Greek *bradus* [slow] and *seismos* [movement]). It is now known that such variations are common in volcanic areas, but there is no place on Earth where they have been so obvious for such a long time. When magma withdrew deeper into the crust, the land would sink at Pozzuoli. On the other hand, when the magma started to move upwards again, Pozzuoli would rise. If the magma reaches the surface, it erupts in a volcano. This is what happened in 1538.

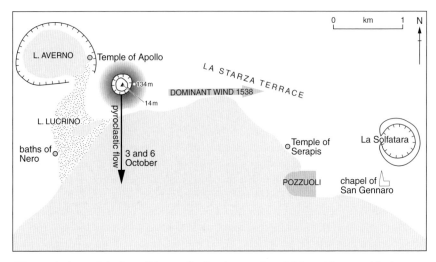

The north shore of the bay of Pozzuoli after the eruption of Monte Nuovo, with the crater of Averno volcano forming a circular hollow now separated from the bay mainly by deposits expelled from Monte Nuovo.

Warnings of an eruption

As the magma was making its way up towards the land surface during the fifteenth and early sixteenth centuries, it caused the Earth's crust to bulge upwards in and around Pozzuoli. The land did not bulge up constantly, like a balloon being blown up. The Earth's crust could resist the upward pressure of the magma for a time, but it would have to give way as the pressure of the rising magma eventually broke its resistance of the Earth's crust. When the crust gave way, the land jerked upwards. Further rises of magma renewed the pressure and caused another jerk upwards, and so on. The process would continue until either the magma stopped rising or it reached the land surface and erupted as a volcano. Each upward jerk of the crust produced an earthquake. The individual movements were very small, but they eventually amounted to a total upward swelling of between 5 m and 8 m. Notable earthquakes took place, for instance, on 25 May 1469, 11 August 1475, 31 July 1488, 9 November 1496, 18 March 1499, 18 May 1505, and on 25 January 1508. The earthquake in 1488 seems to have been quite severe, for buildings were destroyed and several people were killed in Pozzuoli. During the first decades of the sixteenth century, these earthquakes occurred with greater frequency, and they caused more damage and increasing perplexity and fear

The Roman market known as the Temple of Serapis at Pozzuoli.

throughout the district. The people had no inkling of the mechanisms behind these events, nor of what might happen in the future.

The clearest result of all these earthquakes was that the floor of the Temple of Serapis was dry again, and Pozzuoli and its surroundings had risen several metres out of the sea. New land was being exposed all along the coast between Pozzuoli and Baia. In 1501 and 1503, the Spanish monarchy issued edicts granting the emerging shore to the community of Pozzuoli. The exposure of yet more land was confirmed by a new privilege granted to Pozzuoli on 23 May 1511 by the viceroy, Raimundo de Cardona. This was not as generous as it might at first seem: it was to establish ownership of the land, so that the recipients would have to pay taxes upon it. The citizens of Pozzuoli were perhaps seeking some consolation when they founded a new parish church of Santa Maria on the newly exposed territory soon afterwards.

The more the sixteenth century progressed, the more the peace of mind of the inhabitants was disturbed. Cola Aniello Pacca was one of those who recorded of some of the major events. For instance, on Easter Saturday, 4 April 1534, a priest in the cathedral in Naples was apparently reading about the Resurrection from Matthew 28.1, 2:

> In the end of the Sabbath, as it began to dawn toward the first day of the week, came Mary Magdalene and the other Mary to see the sepulchre. And, behold, there was a great earthquake: for the angel of the

Lord descended from heaven, and came and rolled back the stone from the door, and sat upon it.

And behold! at that very moment, Pacca claimed, "an enormous movement rocked the whole city". Pacca also reported that there had been another powerful earthquake "on the eighth day of November in 1534". This happened to be his birthday. Although the shocks seem to have been authentic, Pacca was clearly too fond of coincidences for the dates in his testimony to be accepted unreservedly. Antonio Castaldo, for instance, placed the Easter Saturday earthquake on 20 April 1538.

What is clear from the testimony of several eye witnesses is that the ground was shaking even more often from 1534 onwards, most notably during two particularly severe periods: the first spell lasted throughout the month of May 1537, and the second occurred in the early summer of 1538. The Neapolitan doctor and philosopher, Simone Porzio, corroborated this testimony. There had, he said, been strong earthquakes in and around Pozzuoli for two years before the eruption, and "no house remained undamaged and all the buildings seemed threatened with certain and imminent destruction". Castaldo, too, described how the tremors became increasingly severe during 1538, and were felt both day and night, as the "Sun entered [the constellation of] Libra" and summer turned to autumn.

During 27 and 28 September 1538, the earthquakes reached a climax, rocking and shaking Pozzuoli almost continually. Marco Antonio delli Falconi, a priest in Pozzuoli and later Bishop of Cariati, recalled that many wells in the district had suddenly dried up. This was not then appreciated as a valuable warning that an eruption could be likely, because rising magma often swells the ground and blocks the supply of water to the wells.

Then, at noon on 28 September, the land around Pozzuoli suddenly rose by about 5 m and the sea retreated some 350 m from the former shoreline. Startled flapping fish were suddenly stranded on the sands. The prospects of an unexpected bonanza quelled the fears of the citizens of Pozzuoli, who rushed out to collect them. It was a cruel diversion. In fact, the citizens could not appreciate the danger they were in; but they might, perhaps, have taken greater heed of the obvious warnings when the ground started to crack open and give off water and fumes. On the morning of 29 September, according to Francesco del Nero, clear cold water issued from some of these new fissures, and tepid sulphurous water emerged from others. And the ground was probably shaking even more powerfully in nearby Tripergole.

The eruption begins: Sunday 29 September 1538

The eruption began on the "Feast of Saint Michael the Archangel", on Sunday 29 September. According to Delli Falconi, an hour or two after sunset, molten rock burst out of a "hideous abyss" that had cracked the ground open near the sweating rooms in the spa at Tripergole. Russo said that the eruption had started near the hospital in the Fumosa Boulevard not far from the sea. The fissure lengthened through the town as if the ground had been unzipped. Loud explosions followed, which resounded as far away as Naples. The main vent probably developed Tripergole itself and shot glowing red-hot fragments of all sizes into the sky. They showered down on Tripergole and within a few hours must have buried the baths and every building in the village. No records of the crisis have survived from Tripergole itself, but the commentators from Pozzuoli make no mention of any deaths in the spa. It seems more than likely, then, that the sick, the infirm, the "people with money to spend" and all the inhabitants had abandoned the town as the earthquakes had become more severe during the previous two days.

Watching from Pozzuoli, Francesco del Nero was terrified, and yet drawn to the scene:

> The fire appeared with so much force, noise and splendour that I was much afraid in my garden, but not so much that, after 40 minutes had passed, I went to a nearby hill from which I could see everything, although I was half sick. 'Pon my word it was a fine fire.

In Pozzuoli, too, the Neapolitan doctor, Pietro Giacomo Toleto, had similar feelings. In fact, for a small eruption, it was just the right size to create the maximum effect. It was large enough to provide a superb spectacle (especially at night), but small enough to be approached so closely that most of the details of the events could be clearly observed. In contrast, most of the eruptions of Vesuvius had to be watched at a respectful distance, from where many eruptive details were obscured:

> The land opened . . . and formed a fearsome mouth that violently vomited smoke, fire, stones and a kind of mud formed from the ash. And, at the same time as the explosions, there was a noise that resembled a clap of thunder. The fire that issued from this mouth made its way towards the walls of this unfortunate city [Pozzuoli]. The smoke was

partly black and partly white. The first was darker than the darkness of the night and the second was as white as the finest cotton. The smoke that rose into the air seemed to reach the very vault of heaven.

Whatever Toleto believed the height of the vault of heaven to be, it is unlikely that the plume of fumes rose much more than 3 km into the air. But it was enough to strike fear into the hearts of every onlooker, for they had certainly never seen anything like it in their lives. They tried to make comparisons with familiar phenomena, probably only to bring the events before them into perspective, but perhaps also to make them seem more normal, less fearsome and less diabolical.

From time to time, Toleto and his fellow spectators were astonished to see blocks "larger than an ox" being fired from the mouth. "Sometimes they were thrown as far as the distance of a crossbow shot, but at other times they dropped back into the mouth of the mountain". The "mouth" was giving off a strong smell of sulphur compounds. "It was as if the fragments had been thrown from a piece of artillery and had passed through the smoke of gunpowder that had been set alight". The wind carried the finer fragments towards Pozzuoli and eventually as far as Apulia and Calabria. Toleto described how "an infinite number of birds and animals of different species, which had been covered by this sulphurous ash, fell dead or spontaneously allowed themselves to be taken prisoner in the hands of man".

However, the damage created was greater than could have been caused by falling ash alone. It is very likely that at least one pyroclastic flow surged towards Pozzuoli. Many trees, shrubs, and vines were knocked over and stripped of their leaves and branches, and they were damaged to such an extent that it was often impossible to tell what kind of trees they were. The explosions, too, were unusually loud. As Toleto reported, "when this eruption was in its most violent phases, a huge noise could be heard in Naples – similar to a clap of thunder, like that of a vigorous artillery when two armies have just joined battle". The ash and cinders were also notably muddy, especially when they fell some distance from the vent. Even as far away as Naples, they stuck to the facades of the palaces, and "quite spoilt their beauty". The explanation for these unusual features is that this was a hydrovolcanic eruption, where groundwater had mixed with the magma as it reached the surface. This mixture forms the steam that makes the explosions more vigorous, often expelling large blocks and masses of muddy fragments as well as small pyroclastic flows. Such eruptions also build notably squat cones, which have

Monte Nuovo from Lake Averno, with the so-called Temple of Apollo on the lake shore to the left.

a broad base in relation to their height and a very large, deep crater. Monte Nuovo has exactly this form.

The effects of the eruption on Pozzuoli

Monte Nuovo erupted after the ground had shaken almost continuously for two days and damaged most of the buildings in Pozzuoli. The earthquakes had built up the fears of the citizens just as surely as they had knocked down their homes. The eruption brought their terror to new heights. The earthquakes removed any notion that they could barricade themselves in their homes, as many Pompeiians had done, but on the other hand the ash was not falling thickly enough to discourage a mass exodus. Those who had not panicked and left home when the earthquakes reached their climax, panicked and left their homes as soon as the ash began to rain down on the city.

As Delli Falconi recounted:

> The poor citizens of Pozzuoli were terrified by such a horrible spectacle. They abandoned their houses, which had been filled by the ashy and muddy rain that had continued over the whole area throughout the day. Fleeing from death, but with the colours of death painted on their faces, they guided their terrified families towards Naples, some with their children in their arms, some with sacks filled with their possessions, others leading a laden donkey. Yet others were carrying birds of all sorts, which had fallen dead in great numbers when the eruption began. Others took away stranded fish, which they had found a-plenty on the dried-up shore.

Even 50 years later, Russo recalled how some people had fled naked into the night. As he and his family were leaving Pozzuoli, they saw "Zizula, the wife of Geronimo Barbiero, who had gone out in a shirt, mounted on a horse, riding astride like a man. It seemed as if the world was coming to an end, and everyone was weeping and wailing and crying out 'The Lord have mercy upon us!'."

The refugees rushed about the countryside, without knowing where they were running, until exhaustion brought them to their senses. As is common in such crises, the victims made for the nearest large town. Thus it was that as many as several thousand people rushed, tired and dishevelled, into Naples. Their arrival only increased the panic in the capital, where the explosions were booming forth like gunfire and where muddy ash had splattered the buildings and had already formed a carpet, 2 cm deep, in the streets. As Francesco Marchesino wrote, "it seemed like January, and all Naples and its neighbourhood were covered [like snow] in the same dust". It seemed to Marchesino, who had a religious turn of mind, that "God was recalling the first day of Lent, and reminding Man that he must die".

The eruption soon stirred the viceroy, Pedro de Toledo, into action. On Monday 30 September, he rode out with a vast retinue to inspect the new phenomenon and to assess the situation in the Campi Flegrei. They managed to ride as far as the chapel commemorating the martyrdom of San Gennaro, near La Solfatara. They would have had a fine view of the devastated area to the west, had it not been for the strong winds blowing clouds of ash straight into their faces. Scipione Miccio, the viceroy's biographer, described the poignant scene where "they looked down upon the terrifying spectacle and the wretched city [Pozzuoli], which was so completely covered with ash that the remains of the houses could scarcely be seen".

Pedro de Toledo was more used to commanding men than constraining nature. He was realistic enough to know when he was powerless to act. After a while, the viceregal party turned back to Naples, mud-bespattered, and perhaps not much wiser. That evening, someone in the Spanish hierarchy suggested that there could be a hidden meaning in these dramatic events. Del Nero recounted how some believed that there might be a portent in the eruption. Because the volcanic projectiles had been fired from west to east, they guessed that the eruption indicated that Emperor Charles V would soon be attacking the Turks. In fact, the news had not yet reached Naples that the Turks, under the renegade Barbarossa, had already trounced the Imperial forces at Prévesa, in western Greece, on 27 September.

The viceroy had been perfectly correct in keeping a safe distance from Monte Nuovo. The volcano was making a fearsome noise, and, more importantly, it was still expelling thick clouds of wet ash. Such was the power of the eruption that Toleto and Simone Porzio both estimated that a mountain 1000 paces (1850 m) high had been formed within less than 12 hours (Vesuvius itself rose to less than 1200 m at the time.). Delli Falconi commented that "it might seem incredible to someone who has never seen it, that such a mountain had been made in a day and a night". The cone of fragments did pile up quite quickly, but calmer surveyors later ascertained that Monte Nuovo reached a height of only 134 m. These exaggerations clearly show the effect of a fearsome natural event upon the psyche of the observers.

It seems, in fact, that activity probably began to calm down soon after the viceroy turned back towards Naples. Marco Antonio delli Falconi thus decided to inspect the eruption and the new marvel that it had created. On his way from Pozzuoli, Delli Falconi met "that honourable and incomparable gentleman, Fabritio Maramaldo", and together they "saw the fire and the many marvellous effects that took place thereafter". However, the eruption was still too vigorous for them to consider climbing the new mountain, for "it was still expelling ash and large stones, with a great din and noise that was even louder than most heavy artillery could make". Nevertheless, the pair enjoyed a good view of the new mountain's surroundings:

> Towards Baia, the sea had retreated a great distance, although it was covered with ash and broken pumice stones. Near the shore it was covered so much that it all seemed dry. Two springs could be seen that had recently been covered with fragments: one, of warm, salty water was just in front of what had been the Queen's House; the other, on the beach about 250 paces [460 m] nearer the fire was giving off cold fresh water [These were perhaps the same springs that had gushed forth just before the eruption began].

Delli Falconi was a well read Renaissance priest, whose thoughts turned to the legends that the poets of antiquity had invoked to account for the eruptions in southern Italy. Such great events needed giants, and, of course, the poets produced them:

> It seemed to me that Typhoeus and Enceladus from Ischia and Etna [respectively], together with innumerable giants, and those from the

Campi Flegrei themselves . . . were come to wage war again with Jupiter . . . Methought I saw those torrents of burning smoke that Pindar
describes in an eruption of Etna.

Delli Falconi's allusion was in fact not correct because Pindar had made a
brief reference in the first Pythian odes to an eruption of lava flows on Etna,
whereas the hydrovolcanic eruption of Monte Nuovo produced none at all.
However, in his Renaissance way, Delli Falconi did attempt to rationalize
the legends of the giants. He suggested that "the wise poets meant no more
by giants than the exhalations, which, shut up in the bowels of the Earth, and
not finding a free passage, open one by their own force and impulse, and
form mountains, just as those have done that occasioned this eruption".

A calm interlude: Tuesday 1 October to Thursday 3 October

Two days of vigorous explosions had formed the bulk of Monte Nuovo by
the time the eruption calmed down on Tuesday 1 October. Francesco del
Nero was one of those who could not conceive that so much volcanic
material had emerged "from that abyss" in such a short space of time and had
covered a vast area that stretched as far away as Eboli. And, the fragments
were choking the shore to such an extent that it looked like a ploughed field.

Once the eruption calmed down, people found another cause for con-
cern, notably when they reverted to a more medieval turn of mind. It was
well established that the "fires" could spread underground and pop up from
new vents at some distance from the original eruption. "There are many fine
disputes among the worthiest men", wrote del Nero, "and some think that
Naples itself is in danger. There have been religious processions. An infinite
number of wells have been dug between Naples and Pozzuoli to extinguish
the fire." Some, indeed, did turn to religion, which was certainly as useful
as digging wells. A solemn procession of the faithful had brought the head
of San Gennaro from Naples to the chapel commemorating his martyrdom
near La Solfatara. When Monte Nuovo quietened down, they naturally concluded that the saintly relic had appeased these new volcanic forces, just as
it had always appeased Vesuvius in the past. Whatever the merits of San Gennaro, and in spite of appearances to the contrary, the activity was not yet
over. But, at least, it was possible to take advantage of the lull.

The mountain did not erupt on Wednesday 2 October. Instead, the pall

The deep crater of Monte Nuovo. The vegetation shows that little activity has taken place since the eruption ceased in October 1538. The flat top of Cape Miseno rises in the distance.

of fumes dissipated and revealed Monte Nuovo to the astonished gaze of hundreds of spectators. Marco Antonio delli Falconi took a boat past the mountain to attend to some business on the island of Ischia at the western end of the bay. No doubt imbued with greater scientific curiosity, Pietro Giacomo Toleto was one of a small party that day who dared climb up to the very summit of the cone. They had not fully apprehended how a volcanic eruption worked, and how craters were formed when fragments were ejected. They were therefore astonished to see that there was a large hole in the core of the mountain. Some compared it to a bowl; others affirmed that it was even larger than the market place in Naples, which was about 150 m across. Toleto "saw that the mouth had a circular concavity, about a quarter of an Italian mile [c. 500 m] in circumference. In its midst, the stones that had fallen back into it were boiling, just like water in a great cauldron that has been placed on the fire". In fact, the crater was 420 m in diameter and 120 m deep. It was much larger and deeper than the craters on cinder cones formed by moderate effusive–explosive eruptions. What had seemed to be boiling material was most probably bubbling molten rock. It was a warning that Monte Nuovo was still active.

Thursday afternoon, 3 October

On the morning of Thursday 3 October, the new cone still appeared silent and tranquil. In the afternoon, Marco Antonio delli Falconi set off to sail back from Ischia. The boat rounded Cape Miseno and turned northeast-wards towards the ash-covered city of Pozzuoli. About 7 km away on the northern horizon, Monte Nuovo now formed a low cone of bare yellow ash. In the afternoon heat, each gust of wind sent up swirling columns of fine dust that twisted across the stark devastated area extending to Pozzuoli.

Suddenly, at about four o'clock, Monte Nuovo exploded again. "A great fire appeared," wrote Delli Falconi:

> An infinite number of billowing columns of smoke shot up within a very short time, with the loudest noise that I had ever heard . . . So much smoke expanded over the sea that it came close to our boat, which was more than 4 Italian miles [c. 8 km] from the place of their birth. This mountain of ash, stones and smoke seemed as if it was going to cover all the sea and the Earth.

The explosion that burst obliquely from Monte Nuovo had all the characteristics of a pyroclastic flow. At that moment, the passengers on the boat were in great danger. The fearsome cloud lost its impetus only a short distance from the vessel – just before it could smash it to bits and send it to the bottom of the sea. Delli Falconi could not appreciate how lucky he had been.

Marchesino explores: Friday 4 October

The eruption on Thursday afternoon probably lasted for no more than a few minutes. No further activity occurred that evening, nor on Friday or Saturday. This interlude gave Francesco Marchesino the chance to leave Naples and explore the new mountain and the devastation around it. Other than that he did not own a horse, nothing is known about him apart from the letter that he wrote describing his experiences to a member of a religious order in Rome. His reactions to the events are closer to long-established religious views than are other accounts. The horseless Marchesino took a boat to Pozzuoli on Friday 4 October. "I was shocked, incredulous and terrified by everything that I saw." It was hard to land. A layer of ash and pumice, 45 cm thick, cluttered the harbour and blanketed the shore all the way past Monte Nuovo and out as far west as Baia. Inland, a mantle of ash had devastated the fields and gardens, and the trees had been knocked down. A shroud of yellowish ash covered the shattered remains of Pozzuoli. He walked up towards the centre of the town. It was completely silent and "without a single citizen . . . There were not ten houses still intact; most were crushed or ruined . . . and in some places hardly one stone remained standing above another." Half of the cathedral had fallen down; the other half was tottering. The building housing the church archives had been burned down. Everyone had abandoned the town at the beginning of the week. The horrified visitor felt that the very buildings were crying out their distress. The scriptures came to mind: 'I am black but comely, o ye daughters of Jerusalem' [Song of Solomon 1.5]; and 'There is no soundness in my flesh because of thine anger, neither is there any rest in my bones because of my sin.' [Psalm 38.3]

But even the scriptures provided no consolation for the desolation of Pozzuoli. Marchesino could bear the sight no longer. He returned to his boat and set out across the bay to Monte Nuovo through the mantle of floating pumice and choking ash. He decided to climb up to the summit:

> The path up the mountain was steep and burning [still hot?], although it was easy to climb . . . [but] as soon as we left the path, we could not go on because of the fire underneath the powdery surface . . . Many small fumes issued little by little from the mountain, just like burning rubbish goes on smouldering after the fire has died down, and only reveals itself by the smoke that it gives off.

At the crest of the mountain, Marchesino was as amazed as Toleto had been to see that its core was hollow. He struggled to describe it. "I found that it was not full inside, but empty . . . it was like a chalice that was wide at the top and narrower lower down. Everything sloped down inside [of the cone] in the same way as it did on the outside . . . it narrows as it descends to its base, which seems like a narrow point without a cavern. It did not seem to descend as much on the inside as it did on the outside." The present base of the crater lies about 12 m higher than the outer foot of the cone.

Marchesino did not mention any of the "boiling" in the crater that had impressed Toleto two days before. It suggests that, by Friday, the eruption was diminishing in vigour and the magma had withdrawn farther down the vent. These conclusions were not to prove entirely justified.

Marchesino was afraid to stay too long at the summit. The cone was sterile, the land around was devastated and deserted. All trace of Tripergole had disappeared except for the old Roman baths then known as the Temple of Apollo. When he returned to Naples that evening, his overwhelming feelings were still of horror and confusion. He set out to write an account of his extraordinary journey to an unknown churchman in Rome, but he felt bound to confess that logic had been driven from his mind: "if this letter seems long, disordered and confused, it is because I have become confused myself and cannot write with the lucidity required". Marchesino then stepped out of history. Monte Nuovo appeared to be about to do the same, for it gave off only a few fumes on Saturday.

Sunday 6 October

The tranquillity of the previous two days had instilled a false sense of security into the population, and especially into those who could not resist a trip up to the summit of the latest local attraction. By the afternoon of Sunday 6 October, several dozen, perhaps even over a hundred, people were making their laborious way up the ash-strewn flanks of the cone.

Suddenly, Monte Nuovo sprang back to life. An explosion ripped out of the crater and a hot ash-laden cloud careered down the southern flanks of the cone. The volcano had unleashed yet another pyroclastic flow. It was by far the most lethal, for its victims were much closer to its source and thus much more vulnerable to its impact. Pietro Toleto's account betrayed his medical interests. "Some", he wrote, "were suffocated by the fumes, some were knocked down by the stones, and some were burned by the flames". All these are characteristic ways in which a pyroclastic flow kills.

Delli Falconi described their fate:

Some people had climbed half way up the mountain, and others had reached a little farther up, when, about the twenty-second hour [4 p.m.], such a fearsome and sudden conflagration arose with such a great mass of smoke, that many of these people were suffocated; and many were never found, dead or alive. I was told that those who were killed, and those that were not found at all, amounted to more than 24 in number. There have been no notable effects since then, but the activity seems to come back from time to time like the ague and the gout.

Delli Falconi was a priest who had received training in both philosophy and the natural sciences, and his enquiring mind led him to speculate about the causes of these great and disturbing events. He believed that "the earth-

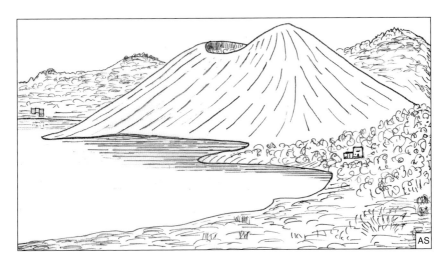

Monte Nuovo in the early nineteenth century, based on an engraving in Lyell's Principles of geology.

quakes, the fires, the drying up of the sea, the amounts of fish and dead birds, the newly formed springs, the rain of ash, both with and without water, and the innumerable trees . . . that had been uprooted and covered in ash . . . had all been produced by the same cause that creates earthquakes". All these features were indeed connected, but not, as many still believed at the time, by underground caverns.

The aftermath

After its burst of temper on 6 October, Monte Nuovo quickly became dormant, although it was not as obvious at the time as it is nowadays with over 400 years of hindsight.

The first reaction of the citizens of Pozzuoli was to abandon their homes and go and live somewhere else. The viceroy, Pedro de Toledo, would have none of this. He was deeply interested in urban development, and he must have jumped at the chance to build on a site such as Pozzuoli, which had lost most of its viable buildings and all of its citizens. Two papal decrees, issued in 1543 and in 1552, helped his task because they exempted the clergy from the *decimo* tax for two periods of ten years. Pedro de Toledo's biographer, Scipione Miccio summarized his efforts to rebuild Pozzuoli:

The viceroy, who was unwilling to consent to the desolation of such an ancient city that was of such use to the world, decreed that all the citizens should be repatriated and exempted from taxes for many years. To demonstrate his good faith, he himself built a palace with a fine strong tower, and erected public fountains and a terrace a mile long with many gardens and springs. He reconstructed the road to Naples and widened the tunnel so that it could be traversed without lights. He built the church of San Francesco at his own expense. He also had the satisfaction of completing his own palace, and of seeing that many Neapolitan gentlemen had built mansions there too. He also restored the hot baths as successfully as possible, and had the city walls rebuilt. And, to stimulate interest in the city, he decided to spend half the year in Pozzuoli, although ill health subsequently enabled him to stay there only in the spring.

Not everyone had the same determination, the executive power, or the financial resources as the viceroy. For months after the eruption, many people remained homeless, partly because their property had been damaged beyond repair, and partly because they were afraid that the eruption and especially the earthquakes would resume. For years, the countryside suffered hardship as the farmers struggled to reclaim their land and waited until weathering made the ash fertile enough to grow crops again. However, as the magma below the surface retreated once more to the depths, earthquakes continued to shake the Campi Flegrei and Pozzuoli on and off for more than 50 years. Monte Nuovo gave off steam and fumes for several centuries. Lazzaro Spallanzani, for instance, saw weak hot fumes emerging from the crater when he climbed the cone in 1794. After a couple of centuries, however, the mountain gradually acquired a cover of vegetation, which now makes the cone more attractive to the visitors to the nature park that has been set up to preserve its environment.

Tripergole has completely vanished, although it is said that fragments of the village can be identified among the ash exposed on Monte Nuovo.

Further reading

Di Vito et al. 1987; Dvorak & Gasperini 1991; Parascandola 1946; Scarth 1999.

Chapter 7

The eruption in 1631: the Counter Reformation

The Church and the Establishment used the violent eruption of 1631 to strengthen their hold on the people by claiming that God had wreaked this death and devastation upon them as a punishment for their sins.

In January 1631, when Manuel de Fonseca y Zuñiga, Count of Monterrey, arrived in Naples as the new viceroy representing King Philip IV of Spain, he cannot have imagined that, before the year was out, he would be faced with the worst volcanic disaster in Europe for over a thousand years. He could have handled a religious or political problem much more readily, for he was very well connected: his brother-in-law was the Count–Duke of Olivares, the real master of Spain.

The wages of sin

In 1631, religion had been the main cause of controversy and war throughout the century since the theological earthquake of the Reformation. The Council of Trent (1545–1563), had tried to stifle the intellectual searching of the Renaissance and had set out to re-establish the orthodox beliefs and revealed truths of the Catholic Church, which had prevailed during the Middle Ages. The Counter Reformation was now in full swing in southern Europe. The Catholic states were in the midst of a war against those presumptuous and heretical princelings of northern Europe. In Spain and its dominions, the zealous servants of the reinvigorated Church were in the forefront of the struggle to ensure the unqualified acceptance of the truths about all aspects of life on Earth that the Almighty had revealed to his representative in Rome. The Inquisition strove assiduously to seek out and condemn the heretics who contested these truths. In Naples, the only saving grace in the prevailing climate was that the Inquisition had never been allowed to enter its gates. Nevertheless, the Church seized eagerly upon any social or natural crisis to bring the people to heel and demonstrate that God was displeased. The cause of such calamities was clear; the people had sinned. Only repentance could placate the Divine anger, and, the more grandiose, theatrical and public this repentance was, then the more likely it was that God would relent. The eruption of Vesuvius in 1631 provided a wonderful opportunity to reassert orthodox doctrines. The leaden hand of the Counter Reformation lay so heavily over all scientific enquiry, which had blossomed since the Renaissance, that it was practically impossible for anyone to propose a rational explanation for the eruption or to organize a rational reaction to it. Fear of accusations of heresy and being burned at the stake was such that all those who dared to disagree with the revealed truth of the Church would be well advised to keep their views to themselves and join the processions of flagellants. The way in which the unprecedented crisis was handled gives an insight into some of the social impacts of the Counter Reformation. Thus, Vesuvius played a major role in the Counter Reformation in Italy.

The events of 1631 also provide an insight into a long-running volcanological controversy about what exactly erupted. The catastrophe gave rise to more than 200 accounts of the events, whose very number at least illustrates a Renaissance attitude to natural phenomena. However, like their predecessors, the writers had never seen an eruption, did not understand, could not analyze, and did not even have the vocabulary to describe in any coherent way what happened. The term "volcano" was still not in common usage, and the volcanic sense of the term "lava" was not to gain currency for another century. As a result, until very recently, the events of 1631 have been misinterpreted perhaps more seriously than those of any other great eruption in the history of Vesuvius. It was long believed that unusually fast-flowing lava flows and mudflows had caused all the devastation, but the recent study of the actual deposits left by the eruption, and a more detailed analysis of the contemporary documents, has made it clear that Vesuvius probably gave off no lava flows at all in 1631.

The eye-witness reports of the eruption come from letters and pamphlets published in Naples immediately after the catastrophe. These accounts sprang both from the Italian establishment and from the Spanish nobility. They all had a vested interest in fostering the view that the establishment had

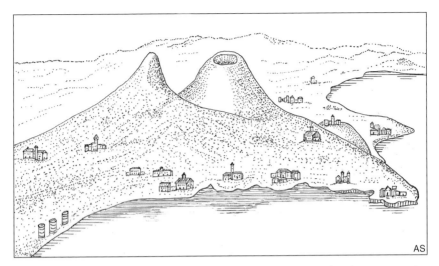

Vesuvius and the ridge of Monte Somma before the eruption in 1631, based on an engraving by Perrey.

saved Naples from the wrath of the Lord when He had caused Vesuvius to erupt. They upheld the status quo, praised the Spanish viceroy, and glorified the role of the Italian religious hierarchy, who had begged the local saints to protect the city and had urged the sinful multitudes to repent. And the piety and superstitions of the ordinary citizens supported this religious status quo almost without a second thought.

Vesuvius in 1631

There is nothing like a catastrophe for developing nostalgia among the survivors. In their mind's eye, what was devastated had been a land of milk and honey. Reality, for once, almost matched the nostalgia. The region had made great economic advances during the previous two centuries. One commentator, Frat'Angelo, described the Campania of yesteryear in words that Verdi might later have set to a resounding chorus: "Tutta arbustata, vitata, olivetata, copioso di vini pretiosissimi, frutti bellissimi, et pascoli abondantissimi" ("replete with shrubs, vineyards, olive trees, full of invaluable wines, superb fruit, and most abundant pastures"). The opulent diversity of the fields matched the wealth of the towns, where the nobility had built ostentatious mansions. But it was often forgotten that the eruptions in the past had punctuated the Campanian idyll with days of devastation. Now and again, earthquakes would still shake the ground, and the citizens would tremble just for a moment at their own vulnerability, until the glamour of their surroundings induced a reassuring calm.

At the beginning of the seventeenth century, the graceful cone of Vesuvius rose some 60–70 m above the protective ridge of Monte Somma to a height of about 1200 m above sea level. Between them lay the grassy vale of the Atrio del Cavallo, at a height of about 450 m, where herdsmen used to pasture their animals, and where apothecaries and the more athletic members of the religious orders searched for medicinal plants. Rough ash and brown stones shrouded the stark crest of the cone of Vesuvius itself. The view from the summit encompassed the scintillating Bay of Naples – breathtaking.

Breathtaking, too, was the huge bowl of the old volcanic crater that sank deep into the cone. A path twisted down onto a central plain, which made a sheltered private world where men pastured their animals and collected wood from the flourishing evergreen oaks, hornbeams and ash trees.

In 1592, the 20-year-old Abbé Braccini and some of his friends had ventured into the sinister cavern in the centre of the crater, but had not dared investigate further. Other explorers had been less inhibited. One fine morning in May 1619, two monks and Domenico Magliocco, the best doctor in Naples, had scrambled on all fours through the boulders into the inner cavern, covered with black sand, from which lava had last emerged in 1139. They found three small basins: "one was full of salty water; one was full of sour, tepid water; and the third contained water that tasted like a chicken broth that had been cooked without salt". They had also felt cold fumes hissing from small holes amid the boulders. Nevertheless, in the early years of the seventeenth century, Vesuvius was hardly displaying the symptoms of an imminent eruption. However, a folk memory of the ancient eruptions persisted among the local farmers, who claimed that a metal gate used to be hidden at the bottom of the inner cavern, and that animals died whenever they strayed past it. It was a harmless enough fable – always provided that Vesuvius was going to remain dormant.

Real, unrecognized and imaginary warnings from Vesuvius

Just after the eruption, many claimed that the eruption of Vesuvius had been the work of the Devil, who, it was said, had issued warnings of his intentions through the lunatics whom he had possessed. Thus, on Saturday, 13 December 1631, a group of gentlemen were laughing and joking in a Neapolitan street when a woman "possessed of the Devil" passed by. "You can laugh today, gentlemen", she exclaimed, "but we shall be the ones who will be laughing on Tuesday." On that same day in the Carmine church in Naples, a man similarly possessed of the Devil was given a drink of water from the very vase that Mary Magdalene was believed to have used to anoint the feet of Christ. The possessed man declared: "I'll make you all repent on Tuesday. I know exactly what I have to do." Curiously enough, many people later tended to discount the role of the Devil, no doubt because they followed the establishment view that God had caused the eruption in order to punish the local sinners.

Vesuvius issued altogether more mundane – and what should have been more frightening – warnings that it was about to spring back into action. These warnings went unrecognized and unheeded. For instance, a man from Ottaviano witnessed an important change within the crater without realizing the implication that magma was rising in the vent. On 10 November 1631,

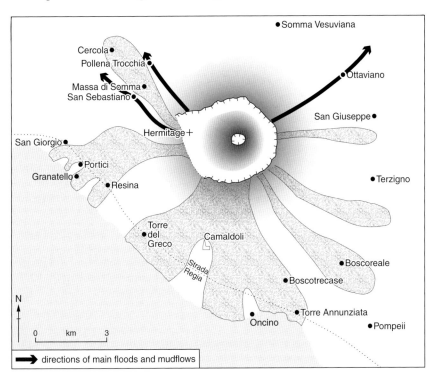

The distribution of pyroclastic flows and the course of floods and mudflows during the eruption of 1631.

139

he had seen nothing untoward. Two weeks later, he was astonished to see that the crater had been filled up to the very brim with "ash and stones", so that he had been able to walk across the crater from one rim to another.

Meanwhile, the inhabitants of several hamlets on the lower slopes of Vesuvius felt an increasing number of minor earthquakes, which could have been caused by magma rising into the volcano. On 10 December, the mountain started rumbling so much that the people in Torre del Greco, in the south, and in Massa di Somma and Pollena, in the north, had scarcely been able to sleep. Francesco Ceraso later wrote that "it seemed as if all the forges in the world had started operating at the same time". Moreover, some wells had dried up, and others had become dirty and muddy, although there had been no rain. People became more anxious when some pundits claimed that the subterranean spirits had been aroused. There was no need for spirits: the natural course of events was quite enough. In the villages around the mountain, many began to beg for the Lord to have mercy upon them.

The eruption begins: Tuesday 16 December

Throughout the night of 15–16 December, the mountain shook and rumbled with a noise like drumrolls and thunder. Almost continuous earthquakes

---

The course of the eruption in 1631

After 492 years of repose and a month of unrecognized warnings, Vesuvius burst into eruption soon after 7 a.m. on Tuesday 16 December 1631. For the rest of that day, a Plinian eruption column soared some 21 km skywards, and the whole area around Vesuvius shook almost continuously. Ash and rain poured down all around the volcano. The wind blew the upper parts of the column eastwards so that ash soon began to fall that evening over southeastern Italy.

At about 10 a.m. on Wednesday 17 December, loud explosions heralded the collapse of the crest of the mountain. At once, pyroclastic flows poured up out of the enlarged crater, swept down the southern flanks of Vesuvius and devastated everything in their path. That same day, the erupted steam and violent thunderstorms combined to cause heavy rains that provoked disastrous mudflows and floods on the northern flanks of the mountain and dowsed the other slopes with wet ash.

Activity began to wane within 48 hours of the start of the eruption and, by the end of the week, the eruption had calmed down enough for some rescue operations to begin. Emissions then continued on a much reduced scale until they ended imperceptibly in a whimper sometime during January 1632.

Sketch, based on an engraving by Passeri, showing the rising Plinian column and pyroclastic flows racing down the flanks of Vesuvius on 17 December 1631. One pyroclastic flow almost surrounds the church in the foreground, another reaches Portici in the middle distance, and a third extends to Resina just beyond it. Two further flows have already overwhelmed and set fire to Torre del Greco and Torre Annunziata and form the promontories on the right of the picture. The buildings in the foreground represent Naples.

began shaking Naples, and strange gusts of wind swirled through the streets. Now that Naples was in danger, the Neapolitans took a much more serious view of events. A particularly long and vigorous quake just after midnight woke up the citizens, who dashed to their windows, only to see their equally terrified neighbours staring back at them. Those who kept their presence of mind counted about 18 shocks before dawn.

The start of the eruption seen on the flanks of Vesuvius
The morning of 16 December dawned windless, cloudless, almost spring-like – and too good to be true. The eruption began not from the main crater

on the crest of Vesuvius but from a fissure that split open the southwestern flanks of the cone at a height of about 700 m. The eye witnesses in the vicinity did not linger long.

Just after 7 a.m., five young men from Sant'Anastasia were on the southern flanks of the mountain when the ground suddenly split open about 1 km away. The display that they described was very much like the start of the eruption of Parícutin in Mexico on 20 February 1943:

> On Vesuvius, smoke and fire came out of the cracks making a noise like gunfire or fireworks at a fiesta. The cracks soon lengthened and, at first, the holes were as wide as wine barrels, but they soon became even wider. Then the emissions of smoke joined together into a single cloud with large stones in it, and flashes of lightning came down through it. One stone landed so close that it nearly hit us.

At the crucial moment, a cow-herd guarding his cattle at the foot of the main cone saw two cracks suddenly open in the ground, and then start to give off smoke and fire as they lengthened towards him. He was burned and almost overcome before he managed to run away. Near by, one man was loading his donkey and another was watching over his goats when they suddenly heard "the first thundering from the cruel chasm". Convinced that the whole world was falling in upon them, they abandoned goats and donkey and never stopped running until they reached the main road to Naples.

The start of the eruption seen from Naples

The eruption had been under way for some minutes before it was noticed in Naples. Eye witnesses struggled for adequate comparisons: Frat'Angelo thought that the noise was like a great storm at sea; Francesco Ceraso said it sounded just as if an army had been trapped inside the mountain, beating drums, driving coaches, and firing off guns. Other Neapolitans described "tongues of flame", "snakes of fire", "rumbling like a furnace roaring full blast" and "a noise like heavy artillery" punctuated by huge cannonades. For the imaginative Spaniard, Simon de Ayala, the cloud "looked like serpents, monsters, or lions, and it soon formed a plume of the most beautiful colours . . . lovelier than dawn on the finest morning in June. Many Neapolitan artists went out to make drawings of it".

Giovanni Battista Manso was one of the foremost intellectuals in the capital and a friend of Tasso and Milton. He rushed to his window when the

142

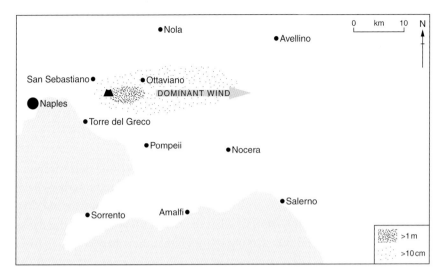

The distribution of airborne fragments erupted in 1631.

opening salvoes of the eruption reverberated through the city around dawn. He saw the erupting column and probably the first pyroclastic flow expelled:

> A column of fire was rushing up above the clouds and a river of flame was spreading down the mountainside . . . It was not really fire, but something half way between flames and smoke, some parts were luminous, but others were darker [a pyroclastic flow] . . . Sometimes flashes of lightning and bigger flames shot out . . . along with a terrible noise like an artillery bombardment . . . [and] an almost continuous shaking of the walls and the ground . . . I thought that I might leave for the country to avoid the damage that seemed imminent, but, when I saw the tumult in the city, I decided not to be the first to set a bad example.

Soon the great column of smoke resembled a gigantic pine tree; blocks were thrown skywards, arrows of lightning shot through the lowering cloud, the mountain rumbled and shook, and explosions blasted from the fissure.

But Vesuvius could do even better than that. About 8 a.m. on 16 December, the summit crater itself began to erupt with a violence it had not displayed for over a thousand years. The column of ash, steam and fumes soon reached the stratosphere and darkened the sky all around the volcano. Large blocks of rock and molten lava exploded from the crater. As the morning progressed, the westerly wind winnowed ash and cinders from the column and scattered them most heavily on Ottaviano and the eastern slopes of Vesuvius. Soon the bright early morning turned into a dark and smoky afternoon as a great mushroom cloud spread a black canopy over the whole volcano and out towards Naples. Darkness began to fall early.

Exodus

By the afternoon of 16 December, the sound of the volcanic bombardment was reverberating throughout Campania. Eye witnesses might perhaps have described the course of events with exaggeration born of their fear, but they surely gave a correct impression of what the victims felt. Around the flanks of Vesuvius, the ground was shaking so much that the villagers rushed from their homes into the safety of the open air, only to be confronted with the fearsome tower of steam and glowing ash issuing from the volcano. Many believed that "the trumpet of the Last Judgment was about to sound". The gravity of the situation notwithstanding, some priests were shocked to see that the women forgot both their dignity and their vital garments, while the men stood gaping at the mountain, paralysed and demoralized with fear.

All around the volcano, the villagers abandoned their homes, their possessions, and sometimes even their families, and ran away as fast as they could. Some carried their children; some just clutched the first things that had come to hand when fear had stricken them. Some men just had time to load their belongings onto their backs, while others dragged away furniture, and even their elderly relatives, on pathetic little carts. They made a pitiful sight: pale, unkempt, and often half-naked and half-burnt, as they rushed headlong towards Naples and the promise of salvation. They could only keep on repeating: "It's raining fire! The Lord have mercy on us! Misericordia! It's raining fire!" No-one could dispute that. The billowing cloud of fumes and ash was already towering several kilometres above Vesuvius. As events turned out, these fugitives had almost certainly saved their own lives, and their blind panic took on an air of prudence after the devastating events that unfolded the following day.

The viceroy acts: 16 December

When the volcanic crisis suddenly loomed on the eastern horizon, the viceroy, the Count of Monterrey, did not betray the sense of duty that he had inherited from a long line of Spanish grandees. He declared that he would stay in Naples, and die with his people if necessary, "rather than flee and leave them behind with neither instructions nor directives". His declaration left the Cardinal–Archbishop of Naples, the head of the religious hierarchy, with no alternative but to vow to die with his flock. Once the twin pinnacles of the establishment had come to their decision, the nobility and the higher clergy were then forced to follow suit.

As soon as the gravity of the eruption became apparent in Naples, the viceroy sent out three trusted men to gather information from three key points in the area: to Pozzuoli in the west, to Capua in the north, and to the flanks of Vesuvius itself in the east. The viceroy also established a public-health commission, which included several doctors, to assess the effects of the eruption on the coastal towns of the bay. In fact, this was not such a scientific act as might at first appear. The Count of Monterrey was not yet particularly concerned about the fate of the people in the countryside: he was worried about Naples. In accordance with the view then current that erupted fumes and ash caused disease, he really wanted to know whether any refugees had been contaminated, and whether they might introduce the plague or other contagious diseases if they were allowed to enter the capital.

The public-health commission duly set off eastwards along the coast road. They found that Resina had already been almost abandoned. Then, as the commissioners approached Torre del Greco, they saw so many people leaving the town that they decided that discretion would be the better part of valour, and they turned back to Naples with the refugees.

Flight from Torre del Greco

Torre del Greco was a pleasant little town situated on the coast, only about 7 km south of the crater. When sulphurous ash and cinders began to rain down onto the town, the volcano seemed to loom even nearer, and many citizens of Torre del Greco decided to dash to Naples, where they hoped the holy relics might save their lives, or at least enable them to die with the consolations of religion. However, others opted to seek spiritual help in the local churches, which had the added advantage of being solidly built in stone.

Unfortunately, two of the high-ranking residents of Torre del Greco – the Cardinal–Archbishop of Naples and the governor of the town – did not behave with the decorum befitting their status. Cardinal Francesco Buoncompagno preferred to live in Torre del Greco because he thought that the air was better for his rather delicate health than the more polluted atmosphere of Naples itself. This self-indulgence nearly cost him his life. On the morning of 16 December, Torre del Greco was certainly no place for a sickly cardinal. The earthquakes had awakened His Eminence, and the first sight of the erupting "flames" convinced him that he should leave for Naples without delay. The earth tremors had already blocked the main door of the palace, and the cardinal and his retinue were forced to scramble, half-dressed, down a high wall. They were then told (wrongly) that the eruption had already cut the road to Naples and salvation, so they rushed to the shore to commandeer a boat. But all the boatmen had taken their vessels safely out to sea, where they had a grandstand view of the increasingly terrifying eruptions from Vesuvius. Neither prayer, pleading, nor promises of ducats could persuade any of these boatmen to return to the shore. Eventually, the cardinal's servants discovered a dilapidated fishing boat that no-one had even bothered to take out to sea. It would have to do. Members of the archbishop's retinue helped two fishermen to row the boat and its flustered cargo away to safety. Thus it was that, a couple of breathless hours later, one of the most powerful prelates in Christendom struggled into the harbour at Naples in a miserable skiff that any discerning goat would have disdained.

Meanwhile, fear had sharpened the wits of Antonio de Luna, the Spanish governor of Torre del Greco, and he hit upon a ruse to escape from the town with a modicum of dignity. He hurried to the prison, rounded up a dozen felons, put them in chains, and declared that they had to be saved so that they could be brought to justice. The governor duly set off with them on the road to Naples, where a flock of frightened citizens soon attached themselves to his procession, trusting that he would guide them to safety. When they reached the outskirts of the capital that evening, they were not well received. Indeed, at first, they were not received at all.

As they neared the Maddalena Bridge at the eastern entrance to Naples, they joined an enormous throng of fugitives. The viceroy had ordered the troops to stop refugees from entering Naples just in case they had been contaminated with disease. However, when the viceroy learned of the multitude clamouring at the Maddalena Bridge, he decided that an imminent riot would be a greater threat to public order than an epidemic. He therefore ordered the bridge to be opened to let the refugees enter the city. Unfortunately, by this time, many fugitives from Torre del Greco had already set off back home to face their fate. As soon as the viceroy received Antonio de Luna, he saw through the governor's blatant initiative and ordered him to return to his post forthwith. At first light on the following morning, Antonio de Luna galloped back to Torre del Greco, where he was to make a more praiseworthy contribution to the fate of the town.

The first religious procession, Tuesday, 16 December

The incessant explosions of Vesuvius made every building in Naples rock, creak and groan. Every citizen trembled. A strong smell of sulphur and pitch spread through the city, which brought hellfire and brimstone to mind. Since the Neapolitans could do nothing to silence the volcano, they turned to appeasement in the only form that they knew. More and more people began to think seriously of repentance. Indeed, as soon as he had recovered his composure after his sea voyage from Torre del Greco, the Cardinal–Archbishop of Naples also concluded that it was time for the people to repent.

In the cathedral, he discovered the happy omen that the blood of the San Gennaro had liquefied. The blood of the saint had always liquefied during crises in the past; and the saint would save Naples now. The cardinal–archbishop begged the viceroy "to join him in the pious work that was indispensable to counteract the evil threatening the city". They decreed that a procession should be organized for 2 p.m. to carry the head of San Gennaro to confront the erupting mountain near the church of the Madonna del Carmine at the eastern end of the city, where the relics of the saint were believed to have first saved the city during the great eruption of about AD 472.

Every important group and personage in the capital joined the solemn procession: the clergy, the monastic orders, the collegiate authorities, and, of course, His Excellency the Viceroy, and his civil and military administration. Sad to say, the zealous cardinal–archbishop proved unable to join them. Just as the procession was about to leave the cathedral, he was overcome by a feverish attack, caused, it was said, either by his arduous voyage across the bay or by his fervent prayers after he had landed in Naples. Cardinal Buoncampagno was 35 years old.

A vast crowd of penitents rushed to join the procession and the most

Vesuvius, the Lord and the Sinners

Perhaps better than any of his contemporaries, the Spanish soldier and grandee, Diego, the Duke of Estrada, summed up the reaction of the Counter Reformation establishment to the great volcanic crisis in Naples. His dramatic account of the eruption offers something of the verbal equivalent of sculptures by Bernini or paintings by El Greco. He combined the unshakable religious convictions of a Spanish aristocrat with obsessions with sex and sin and their role in natural catastrophes. He was no stranger to sin himself, for, in his youth, he had caught his fiancée with a young man in suspicious circumstances and had slain them both on the spot. He eventually retired to Spain, wrote his memoirs, and became a member of the religious order of Saint John of God, the Brothers Hospitallers.

I have heard it said that . . . some had to flee completely naked, others were dressed only in shirts, and some were only half clothed. Honest matrons, naked or otherwise, were clothed with great charity by those who had shown greater forethought and had taken with them other garments and spare clothing. The poor tearful nuns said psalms and prayers, and gave themselves to anyone who wanted to be accompanied without any danger of dishonour or violence . . . No-one wanted to steal anything, because they were all certain that they were about to die . . . Not a single mortal sin was known to have been committed in Naples and its surroundings. It illustrates very well the power of the conflagration and the widespread terror of all the people . . .

The people were so terrified, confused and dumbfounded that they had great difficulty in distinguishing dreams from reality, or in realizing what their eyes were seeing, or what their ears were hearing; because their reason refused to accept it. But it seemed clear to them that this was a real fire like that which had been prophesied that would destroy the world . . . and the people, terrified by such unbridled fury and the continuous earthquakes, lost all hope of survival.

At this point, the lovers of the flesh, the depraved and the concubines abandoned the illicit and lascivious arms of their lovers, and took their leave of them without even bothering to ask for their consent. They cursed the moment that they had entered into such a state and, fearing that an unhappy death would supervene, they began their inevitable penitence without further ado. The arms that had so recently embraced their lovers now stretched out into the shape of a cross and they proffered their once immoral and lascivious kisses to the ground, and begged forgiveness for their sins. So great was this indescribable terror that even conjugal rights were revoked, because the consummation of marriage was considered indecent when so many people were so near to death. And those who were driven by the warmth in their hearts embraced each other as if they wished to hide and find consolation within the arms of a companion. Children thought that they would be safer in the protection of their mothers' arms, and sought solace at the maternal breast so that they could shut out the sight of the fearsome flames, the noisome earthquakes and the clouds of ash. At length, the people left their homes and farms, and fled, shouting, trembling, weeping, fainting, crying out and lamenting their fate. They gave up the respect that they owed to their fathers, to the fidelity of their spouses, to the bonds of blood that they shared with their brothers, or even to the love that they should have given their children. They left their homes and took refuge in the churches, where they believed that they would be in greater safety. There they implored the Lord to have mercy upon them, and some even died after they had confessed their sins . . .

They filled the great squares of Naples, where they believed that they would be safer out in the open, away from the collapsing buildings, if not from the quaking ground . . . In every public square were heard the confessions of prostitutes, assassins and blaspheming robbers, who had made the very ground tremble with the enormity of their unspeakable and sacrilegious sins. They made their confessions in public, without fear of condemnation, yearning for salvation and fearing only the judgement of Heaven. With refined penitence, the nuns and monks emerged from their convents and monasteries to implore His Divine Majesty to pardon the afflicted people.

The establishment seized the chance to foster religious fervour and to strengthen their own power over the afflicted masses. As the Duke of Estrada clearly implied, all this was, as usual, the fault of the underprivileged classes. The poor and the women had been the chief sinners.

enthusiastic walked barefoot in the mud, half-dressed and humble. It was great Neapolitan show of public piety and masochism. Some carried immense crosses and even beams across their shoulders. They mortified their flesh as best they could, lamenting and shouting for the mercy of Heaven, beating each other with their bare hands, with ropes, or iron chains that drew much blood. And yet, for all this display of contrition, the eruption only worsened.

A drumroll

The viceroy inadvertently added to the confusion and panic in Naples during the evening of 16 December. After his return from the procession, His Excellency suffered an attack of religious fervour that stifled his common sense. At about 7 p.m., he issued an order announcing that no-one should have commerce with prostitutes until further notice – on pain of being sent to the galleys. The basis of this decree was that, if the couple were to die during the carnal act, they would be in mortal sin and would suffer eternal damnation. It is just possible that the viceroy was overestimating the carnal appetites of even the Neapolitans in the throes of such a terrifying crisis.

Frat'Angelo reported the unexpected repercussions of this worthy edict. Like all viceregal announcements, the dire news was conveyed to the citizens by twelve town criers, who paraded through Naples banging their drums to attract attention. In the midst of the hullabaloo coming from Vesuvius, many people heard the drumrolls but could not grasp what was being announced. They not only wondered what was going on but concluded that the authorities must be revealing yet another dreadful calamity. Even those who managed to hear the viceregal message were not at all reassured. They reasoned that, if those who were in bed with prostitutes were likely to die, then those who were engaged in more innocent pursuits must be in equal danger of meeting their maker. They, too, were filled with dread. Panic increased through the city. It was not revealed whether those engaged in carnal acts at the time of the edict ever heeded the viceroy's warning, or whether they suffered Divine consequences, although the Duke of Estrada later rejoiced that no mortal sins were committed in Naples at the time.

The night of 16–17 December

Just before sunset on 16 December a cold and dismal deluge of dust, ash and rain began to fall, which lasted throughout the dank and miserable night. The falling fragments muffled the sounds in the streets and extinguished torches, but loud explosions still reverberated all around the volcano. In Naples, the almost continual earthquakes caused even more anguish than the falling ash, for the ground was jumping up and down and from side to side. Many citizens needed little convincing that the Earth was about to swallow them up.

In the countryside, some of those who had not already fled barricaded themselves in their homes, while others shivered in the open air. Some resourceful men built little shacks to shelter their families in the open squares, while richer people tried to sleep in their coaches. Many naturally took refuge in the churches, where the priests seized their chance and preached sermons about repentance, instructed the ignorant, and took confessions galore. Some few, it seems, had earned the repose of the pious. Abbé Braccini, the lucky man, claimed that he had commended his soul to God, the Virgin and the saints, had gone to bed, and slept like a log for seven hours. "These continuous earthquakes", he remarked, "made me feel as if I was being nursed in my cradle, just as I used to be 59 years ago".

News of the eruption probably spread more rapidly by the stratospheric winds than by word of mouth. The upper winds drove the dust and ash towards the east at an average speed of about 54 km an hour. Within eight hours of the start of the eruption, ash was falling over most of southeastern Italy. It started to fall lightly on the Balkans during the following night, and began to fall on Constantinople soon after dawn on 17 December. These sparse ashfalls in the Balkans caused more alarm than danger, but in the immediate neighbourhood of the volcano, crops and livestock soon lay dead beneath an ashen blanket more than a metre deep.

Pyroclastic flows: Wednesday 17 December

During the early morning of Wednesday 17 December, violent explosions from the crater succeeded each other almost every minute, shaking the ground and sending a great cloud of ash and fumes 25 km into the stratosphere, which hid the Sun from most of eastern Campania. Lightning flashed. Torrents of rain deluged the eastern flanks of the mountain.

Ash poured down all over Naples. Religion took over. The churches were crowded. Abbé Braccini declared that "some people went to confession who had never considered confessing before – and one of these was no less than 36 years old". The priests were so overwhelmed that Cardinal Buoncompagno authorized other pious men "of good repute" to take confessions. Abbé Braccini, after his good night's rest, was proud to declare that he had been one of their number. Many confessions had to be taken outside, and some panic-stricken sinners, too desperate to await their turn, shouted out lists of their sins in the public squares for all to hear. Each earthquake, or explosive blast, or even gust of wind, sent yet more people into a repentant frenzy. Many refugees and terrified citizens assembled in the market place, the largest square in the city, or wandered hopelessly about the streets, afraid that their homes would collapse upon them but not otherwise knowing where to turn.

Soon after 10.00 a.m. on 17 December, it seemed indeed that the Last Judgment was about to be delivered. An earthquake shook the city for five minutes that seemed like an eternity. The summit of the cone collapsed into the volcanic vent and pyroclastic flows boiled out from the crater. An old woman in Granatello described how the flow had emerged completely white, "like a silver baton", and had rolled over the ground at first (as a pyroclastic flow). Then a huge cloud of smoke appeared (the pyroclastic surge) that was as white as cotton at its base, darker higher up, and glowing red at its crest. A billowing emulsion of gas, steam, dust, ash and boulders of all shapes and sizes gushed from the crater and spilled down the mountain. Francesco Ceraso described how the pyroclastic flow formed:

> . . . a great smoke, composed of sulphurous ash and other bituminous matters . . . It suddenly blocked the paths of those who were fleeing to safety, ravaged the surrounding area with ruin and death, pulled up trees, demolished buildings, and caused a great massacre of men and animals . . . In the darkness caused by the cloud, most people did not know where to tread.

The cone looked as if it had liquefied. The spectators believed that their last hour had come. Thousands of people around Vesuvius were not mistaken.

During the next two hours, pyroclastic flows gushed down slope in distinct channels. One flowed northwestwards and destroyed most of Massa di Somma, Pollena and San Sebastiano. Another, more than 1 km wide, flowed due westwards towards Granatello and split into four lobes before it reached the sea, destroying the famous pomegranate groves that had given Granatello its name. The largest pyroclastic flow flowed southwards, and divided into two main currents. One attacked Torre del Greco and the other made for Torre Annunziata. Both towns were severely damaged. A smaller pyroclastic flow devastated the village of Boscotrecase.

Where dense and heavily laden pyroclastic flows reached the coast, they formed small peninsulas jutting out into the sea, notably at Granatello, Torre del Greco and Torre Annunziata, where they boiled and grilled many fish that soon began to float dead on the waves. The citizens who had escaped the pyroclastic flows were afraid at first that these fish would be contaminated, but hunger quickly persuaded them otherwise.

Some people who saw the pyroclastic flows approaching from afar escaped from danger if they had the luck, or presence of mind, to run sideways and up hill away from them. But pyroclastic flows can kill by a whole gamut of weapons, ranging from hot ash to gas, rocks and masonry. The commentators described chilling images of the lottery of survival. They seem prone to exaggeration, but, if the flows recently erupted on Unzen in Japan and Montserrat in the West Indies are anything to go by, the commentators probably gave as accurate an impression as any pen could provide at the time.

The pyroclastic flows reach Torre del Greco

Early on the morning of 17 December, Governor Antonio de Luna arrived back in Torre del Greco to resume his responsibilities, no doubt with the stinging viceregal criticisms still ringing in his ears. He tried to bring the situation under what administrators in a crisis call "control". This time, he resolved to do his duty and take charge of his people. They included all those who had made a fruitless journey to Naples on the previous evening; those who had come back to salvage their belongings; those who had returned to search for their lost children; those who had sought refuge in the churches in the town; and those who had cowered, petrified with fear, in the deepest recesses of their homes. A panic-stricken throng of men, women, and children, carts, donkeys, coaches and domestic animals was already cluttering every thoroughfare in the town. Antonio de Luna and several dozen horsemen gathered the crowd together and they were preparing to leave the town by the Naples gate when the first pyroclastic flows foamed out from the summit of Vesuvius and filled the assembly with horror. About 500 people rushed into the churches, and many more tried to run westwards to the protective womb of Naples. They had a few seconds in which to take the only decision that could possibly save their lives. Whatever the governor's plans had been, the arrival of the pyroclastic flow overtook them. With a fearsome noise, the glowing cloud scythed through the multitude, burned at least 2000 people to death and swept many of them into the sea. That afternoon, the waves threw the pathetic remnants of their bodies and possessions back onto the steaming shore. About 150 other people were more fortunate: they survived in the new church of Madonna delle Grazie, because it had been built on a small hill and the pyroclastic flow passed around it. It was said that, until these refugees were rescued several days later, they had been forced to eat the charred bodies of animals that the flow had dumped near by.

Tsunamis

At about 9 a.m. on Wednesday morning, the sea retreated from the shores of the Bay of Naples and returned a few minutes later. The water level fell by between 3 m and 6 m, exposing stretches of coast up to 2 km wide. Ships were left stranded and it seemed as if many of the galleons in Naples harbour would topple over and founder. Ten minutes later, the waters rushed back

The role of the pyroclastic flows in 1631

Fortunately, the pyroclastic flows unleashed by Vesuvius in 1631 were not as powerful as they might have been, for they were denser, moved more slowly, and were less devastating than many of their counterparts. They seem to have gushed out when the summit of the cone collapsed into the vent and greatly enlarged the crater. They probably travelled at speeds ranging between 100 km and 200 km an hour – fast enough to kill, not fast enough to obliterate a whole town. They followed well defined paths down slope under the influence of gravity and the lie of the land. In the towns, for instance, the pattern of the buildings and streets mainly determined their courses; and in the countryside, valleys were more devastated than the adjacent hillsides. For example, the Cappuccini monastery in Torre del Greco, and the church of Santa Maria di Apuliano near Granatello, were both saved because they stood on higher ground, whereas the lower reaches of both towns were devastated. Nevertheless, these pyroclastic flows still had enough power to sweep away houses, and their contents, as well as trees, shrubs, animals and human beings. The eye witnesses invoked comparisons with pitch, quicksilver, bitumen, molten tin, searing ash and blazing water.

These pyroclastic flows had been long mistaken for fast-moving lava flows until that view was rectified in 1993 in one of the first studies of the eruption using modern techniques. In fact, even the eye-witness accounts of the effects of these pyroclastic flows make it amazing that they could ever have been taken for lava flows. The accounts refer, for example, to *fiumi di fuoco*, or *torrenti di cenere infuocata* ("streams of fire", or "torrents of blazing ash"), reaching up to 1 km across, that rushed down the mountain at great speed, with billowing clouds rising above them. Manso, for instance, had described emissions that "were not really fire, but something half way between flames and smoke". Some people were burned to death; others were knocked senseless; some had their clothes burned off while their flesh was undamaged, just as if they had been struck by lightning. Some victims had been beheaded or torn apart.

Lava flows start off as streams of molten rock that soon solidify on their surface and then slow down so that most individuals can easily walk away from them. As many lava flows from Vesuvius in recent centuries have shown, they usually invade, ingest and overwhelm buildings, and might sometimes push them down, but they seldom transport them away. Lava flows can burn living beings, but they never rip off their heads. Indeed, no lava flows known to humankind could have claimed so many victims, maimed so many people, uprooted so many trees, and destroyed so many buildings in such a short time as they are supposed to have done in 1631. Moreover, modern techniques of analysis have yet to reveal the slightest Vesuvian lava flow that could be dated unequivocally to 1631. On the other hand, the death and devastation described are absolutely typical of the work of pyroclastic flows, and they were probably the most lethal volcanic features that had been witnessed on the European continent for over a thousand years. And, luckily for the Campanians, Vesuvius has unleashed none of comparable power since.

furiously in a wave reaching up to 5 m high. It smashed the stranded ships against the Molo jetty in Naples, threw many smaller sailing vessels 2 m or more into the air, and left thousands of dead fish on the shore. Three times the sea retreated, and three times the sea returned in a great wave that flooded onto the coast. Nature had clearly gone mad. These waves were tsunamis, a phenomenon that was of course totally unknown at the time. They could have been generated by earthquakes caused when the crest of the cone collapsed into the volcanic vent, or more probably when pyroclastic flows had crashed into the Bay of Naples.

The procession on Wednesday 17 December

People who had escaped the pyroclastic flows and reached Naples told tales of chaos and rampant death around the mountain that vastly increased the anguish of the Neapolitans and sharpened their desire for salvation through penance. They were powerless to do anything else.

Faced with such extreme danger, the cardinal–archbishop and the viceroy had little option but to continue to put the public weal before their own safety. Once again they processed to save the city. The cardinal rose from his sickbed and ordained that the head and blood of San Gennaro should be taken to confront the raging mountain. In spite of the pouring rain, the

The massive Castel Nuovo, which was the headquarters of the Spanish viceroys in Naples, and behind it on the right, with one façade in shade, stands the royal palace of the Bourbon monarchs. Both residences offered splendid views of the eruptions.

155

procession started out after Vespers. Just as the relics of the saint were leaving the cathedral door, the rain stopped and, to the general astonishment, the Sun came out. "Miracolo!" shouted the throng.

There was more. A vision of San Gennaro, wearing the habits of a pope, appeared above the main door of the cathedral, blessed the people assembled in the square and promptly vanished.

The higher clergy and nobility seem to have done penance by proxy compared with those who donned sack-cloth and ashes, and flagellated themselves as they processed through the streets. Outside the Capuan gate, the cardinal–archbishop displayed the miraculous head and liquefied blood of San Gennaro to the angry volcano. Whereupon, the towering cloud turned downwards as if it was bowing its head to the holy relics. The summit of Vesuvius emerged from the fumes: it was much lower than it had been before the eruption. The multitude cried out that another miracle had occurred. When the procession returned to Naples, the cloud surged up once again, but it was no longer directed towards the city (no doubt because the wind had blown it in a different direction). The religious scholars concluded, however, that God was demonstrating that he had saved Naples through San Gennaro. The citizens showed their gratitude by confessing and taking communion – including, it was underlined, "even the public women".

The processions on Thursday 18 December

By Thursday, 18 December, Naples had become a masochist's paradise. Vesuvius was still demonstrating its power with incessant loud rumblings and boiling sounds. Ash and rain fell so thickly that everything seemed covered in mud; earthquakes were shaking the ground every few minutes; it was cold; and a dank putrid smell pervaded the doom-laden streets. The main procession with the relics of San Gennaro was restricted to the short trip from the cathedral to the church of Santa Maria di Costantinopoli. Perhaps the weather was considered too bad; perhaps the novelty of the processions was wearing off; or perhaps the participants were beginning to doubt their effectiveness. Some religious orders felt that the relics paraded hitherto had not done a sufficiently thorough job. More institutions therefore decided to try their hand. The need for repentance probably still remained, but a competitive element and a Neapolitan need to indulge in theatrical display crept into the crisis. The Franciscans from San Lorenzo put ropes around their necks and they brought out their relics of Saint Anthony of Padua. A sumptuous procession of Jesuits paraded relics of their founder, Saint Ignatius, accompanied by a crowd of brothers wearing sack-cloth, beating their chests with stones, and whipping themselves with such vigour that blood spurted from their flesh. Some screamed in real or imagined agony; others tore out their hair in contrition for sins that they were convinced had brought this calamity upon the city. Not to be outdone, the Theatines, Celestines and Dominicans also processed. Together, they made a throng of over 3000 barefoot penitents, mortifying their flesh, reciting the rosary and carrying torches, crucifixes and sacred bones of the dead. Notable among the penitents was "a squad of about 30 women of ill repute", with close-cropped hair and bare feet, with ropes and chains around their necks, and crowns of thorns on their heads. "They displayed all the signs of true repentance and contrition . . . The spectacle was so moving that many observers wept."

During the early days of the eruption, processions dominated public initiatives, and few practical tasks were undertaken to alleviate the effects of the disaster. There was no plan, and any action was left to individuals. An uncalculated spin-off from these arrangements was that the populace was so thoroughly engrossed in the processions that no displays of public disorder broke out in the city. The administration was absolved from blame, which was placed entirely on the consciences of the multitude of sinners in the city. And, of course, once flagellated, the people had little energy for riots.

The floods at Nola: Thursday 18 December

One of the extraordinary features of the eruption in 1631 was the extensive flooding of the northern flanks of Vesuvius at Marigliano, Pollena Trocchia, San Sebastiano and Ottaviano, and especially the plain of Nola. Many eye witnesses were puzzled and dismayed when the mountain erupted what they considered to be the "unnatural combination" of searing hot ash and water. Various pundits asserted that the water had come either from a blocked underground river or from a bottomless cavern in the mountain, or that it had been sucked up from the Bay of Naples. And naturally some were quick to claim that it was Noah's flood all over again. The reality was rather more ordinary: the floodwaters simply came from torrential downpours after ash and dust coagulated the vapour-laden Plinian cloud that had risen into the stratosphere.

Francesco Ceraso claimed that the flood was 10 m deep and 1500 m wide:

It attacked the land near Nola with such fury that the people had no time to save themselves. Many were caught sleeping in the lower floors of their homes. The flood rose rapidly and forced the people to escape to the upper floors and then to go, naked, onto their rooftops and the higher places of their homes. In fact, no-one was safe because boulders were smashing furiously against the buildings and throwing them down. This caused a great loss of life because the people could no longer escape and they fell easy prey to rapacious death . . . The plaintive voices of the dying carried across the waters. It was firmly believed that the Flood had come once more to the Earth.

It needed a swashbuckling Spanish adventurer and soldier to undertake palliative actions more practical than prayer. Captain Alonso de Contreras left a graphic chapter in his memoirs about his role in Nola, where he was in charge of a company of Spanish infantry:

[On 16 December] the people trembled with fear as the day changed into night, and they then began to abandon the town as the ash continued to fall throughout the day and into the following night. That night was so dreadful that I can well believe that it would have no equal, except in the Day of Judgment itself. Indeed, not only the ash, but also earth and blazing stones fell to the ground, which were like the clinker that blacksmiths throw out of their smithies . . . That night . . . there was a continual earthquake that made 37 houses collapse; and the cypresses and orange trees were torn apart as if by a steel axe. It was dreadful to hear everyone crying out "The Lord have mercy upon us".

Daylight hardly came on the Wednesday [17 December], and the candles had to be kept lit all day. I myself went out into the countryside with a squad of soldiers and brought back seven loads of flour and ordered bread to be baked. This was the way to help those who had lost their homes and were camping out in the open because they were afraid of staying under any roof. There were two convents of nuns in the area who did not wish to leave their abode – although the vicar had given them permission to depart before he had fled himself. The two convents collapsed, but without harming a single nun, because they were all praying . . . inside the church . . .

The soldiers of my company were almost on the point of revolting against me. They held a meeting . . . and they decided . . . to force me to leave the area, because the fire was fast approaching. I found them grouped together in one of the streets [and] . . . I knew what was afoot as soon as I saw them. I called out: "Where are you going, lads?" One of them started to answer: "Sir . . .". But without letting him say any more, I said: "Let all those leave who want to leave! As for myself, I shall not depart until my legs have burned. When things have come to that, our flag will not weigh much, and I will carry it away myself." At this, none of them dared utter another word.

We spent that day in either total darkness or feeble light. There were so many piteous sights of the people left behind that the sight was neither easy to describe nor even to believe: dishevelled women, children running here and there, with none of them knowing where to take shelter as they waited for the natural night to fall. Here two houses were collapsing, and over there another one was on fire. But no-one knew what would be the best place to try and escape to, because you always sank into the burning ash and earth that had fallen that Thursday morning.

Water now joined in the fray, although the fire and ash had by no means stopped raining down. Suddenly the mountain gave birth to a river that was so swollen that the noise alone was enough to cause fear and dread. And one of its branches was coming towards Nola. I took 30 soldiers and local men with picks and shovels and we dug a trench and succeeded in diverting the waters. The waters fell upon two farmsteads and swept them away like ants, along with all their cattle, sheep, and oxen that could not get away in time . . .

On Friday [19 December], the Lord willed that water should fall from the Heavens. It mixed with the earth and the ash and made a cement, which set so firmly that it was impossible to break into it with a pick or a shovel. But there was one consolation: it would give us a means of escape if the fire were to threaten us. On the Saturday, nearly all the barracks collapsed where the company was housed. But the soldiers were unharmed because they . . . had preferred to be exposed to the water and the ash in the square.

[On Sunday 21 December, Captain Contreras received an order from the viceroy to go to Capua.] I therefore left Nola with what I had on my back . . . We were a piteous sight when we reached Capua . . .

as if we had just walked out of Hell. Most of us had no shoes; our clothes were half-burnt and our bodies likewise. We were given hospitality and we recovered our strength during the next week. We celebrated Christmas there, although Vesuvius was still vomiting fire.

Rescue and recovery?

Eventually in Naples, too, Renaissance practicalities began to emerge from the mists of the Counter Reformation. It is possible that the viceroy realized, but could not admit publicly, that the religious processions were proving ineffective. On Thursday, 18 December, therefore, he launched a more practical policy and organized several expeditions through the public-health commission, which were repeated for the best part of the following week, although their most laudable aims proved hard to fulfil. He sent two galleys and several smaller vessels laden with provisions for the stricken areas along the coast at the foot of Vesuvius. He instructed the captains to return with as many survivors as they could embark, and to put guards on all property to prevent theft where the people had been forced to leave all their possessions behind. He particularly recommended that all the children lost in the panic should be retrieved and reunited with their families. At the same time, he sent out post horses to bring back news of the extent of the damage all around Vesuvius, especially since the area north and east of the mountain had been completely cut off from Naples. He even sent soldiers and militiamen to try and discover where "the fire" was coming from, and gave them the impossible task of blocking it with all deliberate speed.

For several days, however, none of the scouting sorties was able to progress beyond the southwestern flanks of Vesuvius. They all reported the same appalling tale. Even those who had escaped the brunt of the pyroclastic flows made a piteous sight – stupefied, wounded by flying objects, burnt and stripped half naked by the swirling searing ash. That Thursday, 18 December, the viceroy ordered Captain Gaspar de Zuñiga and Sub-Lieutenant Alexandro to take four men of the light-horse and assess the situation along the coast. They met with little but horrors within the paths taken by the pyroclastic flows. The ash, often over 2 m deep, was still very hot, and foolhardy individuals who had tried to cross it had sunk through the thin crust and had died a horrible death before their companions could pull them to safety. Near Pietra Blanca, they found a dog eating five half-burnt corpses, but they drove off the animal and buried the bodies out of its reach. Farther on, at the church of the Madonna del Soccorso, they had found another ten half-burnt corpses. They found up to 100 bodies in Granatello, but in the church of Santa Maria di Apuliano, they discovered 53 women at prayer. The captain told them that galleys were available to take them to safety in Naples, but the women would have none of this. They said that they would not leave the church because the Blessed Virgin had saved them when the fire and the smoke had come. During the previous three days they had eaten nothing but bread that five bandits had brought for them. When the viceroy later heard of this unexpected turn of events, he sent Sub-Lieutenant Alexandro back again with a boatload of wine, water, bread and fruit.

Friday 19 December

The rescue operations continued on Friday morning under a clear and windless sky. There were fewer earthquakes; the eruption continued to operate in a lower key; and no new scourge emerged to torment the populace. Meanwhile, yet another relic was brought out to face the volcano: the minor Franciscans processed with the intact body of the Blessed Giacomo della Marca in its crystal sarcophagus. The crowd went to confront the mountain from the Maddalena Bridge, where a Dominican preacher harangued them for their sins. By now, some of the participants were finding that processing was becoming rather tiring, and even Cardinal Buoncompagno had felt unable to walk beyond the city gates. In contrast, however, the chief minister of the kingdom, Carlo Tappia, earned general admiration by continuing on foot, "although he was an old man of 66".

On 19 December, however, at least one expedition reached Torre del Greco. The smell was appalling; "the fire" was still burning in the sea; and dead fish and corpses were poking out of the ash the pyroclastic flows had left.

Frat'Angelo gave an unexpurgated version of how the eruption had affected the human body and, even allowing for poetic licence, his description offers clear proof that pyroclastic flows had been the major cause of death. Neither lava flows nor falling ash could have possibly dismembered and injured the victims in such ways:

Most of the people who had stayed in the neighbourhood were killed, suffocated by the smoke and the ash; some were burned by the fire, some were buried by the buildings, and some were covered with stones that had been shot from the chasm. Those who have been there report an infinite number of dead, desiccated, wizened, stinking, headless, armless, legless, many lying with their feet in the air, but some were turned to Heaven. Corpses were . . . roasted, dried up, stinking, and reduced to ashes. Some had only their intestines burnt; some had only their clothes burnt, but others had their clothes intact. Here was a man's thigh, there the shoulder of an ox, and there a person's trunk . . . An animal's head here, a man's head there; here a dead goat, over there a man's thigh, a shoulder, a foot, a leg . . . A dead man's skin had been removed, and the flesh on his feet was like the soles of shoes.

Small wonder then that, when Vesuvius started to expel thicker fumes that evening, the explorers were relieved to retreat post haste to Naples.

The waning phases of the eruption

Saturday 20 December

The public-health commission continued its work with the help of several hundred labourers, who had been chiefly selected from among the Neapolitan tanners, who enjoyed a high reputation for their strength and courage. They had several pressing tasks: to clear the roads of the thick ash and rubble so that normal communications could be resumed and to bury dozens of corpses at a time for fear that the bodies would spread disease. However, it proved impossible to bury more than a small proportion of the victims.

The burial gangs worked in cold wet weather in the shadow of a volcano that was still giving off ash and fumes, albeit not so violently as before. In spite of the rain, the usual assemblies of the authorities and sinners continued their processions in the muddy Neapolitan streets, although hopes were now beginning to enter many a tortured soul that the end of the crisis may be approaching. They took it as a good sign that Vesuvius was not only much lower than before, but that its roaring crater was much wider, which, they hoped, would allow the eruptions to continue without the huge explosions. Yet, every now and again that evening, vigorous earthquakes would shake the region and send hundreds of people, shouting and wailing, into the streets.

Sunday 21 December

The Sabbath was a day of large and fervent processions. Vesuvius was still threatening and a blustery wind was driving torrents of rain through the streets. But at noon, the sky cleared, the Sun came out, and the great processions took place in relative comfort. In a sign of even greater desperation, even more relics were paraded. The fathers of San Francesco di Paolo processed with the milk of the Mother of God; and the Dominicans took the head of Saint Thomas Aquinas from their church and placed it next to the head of San Gennaro in the cathedral. The religious authorities believed, or at least loudly and incessantly declared, that these two saints had saved Naples from the destruction that had been visited upon the neighbouring towns. (It seemed to matter little that the slopes of Vesuvius had been devastated.) Meanwhile, however, the squads of the public-health commission were more usefully devoting their main energies to opening the road to San Sebastiano and Madonna del Arco, and many other groups were still fully occupied in burying the dead.

Monday 22 December

At about 9 a.m. on Monday 22 December, a sudden salvo burst out from Vesuvius and spread fumes all over the mountain. The head of San Gennaro was duly taken to confront the volcano. The worst seemed to be over when the explosive crisis stopped at about noon; the weather improved and the religious authorities and other benefactors distributed alms. But yet again, public confidence was rather shaken when a huge earthquake rocked the ground at about midnight and smaller tremors followed until daybreak.

Tuesday 23 December

On Tuesday, 23 December, the morning sunshine seemed to have calmed down the volcano. Frat'Angelo was one of many who attributed the improved weather and the appeasement of the volcano to the "marvellous result of the benevolence of God and to the intercession of the Virgin Mary, San Gennaro, and all the other saints to whom Naples was so very much devoted". That afternoon, when the usual procession headed by the viceroy passed in front of the castle, the artillerymen fired several salvoes (under orders) to praise his role in saving the city.

Everyone could see that Church and State had worked in laudable unison. No-one seemed to notice, or could admit to noticing, that their efforts had already taken the best part of a week to take the slightest effect. No-one invoked the role of the wind that had carried dust and ash almost exclusively eastwards, away from Naples; nor the role of the torrential rains that had fallen chiefly on the northeastern slopes of the volcano, nor the role of Monte Somma in protecting the northern flanks of Vesuvius from the pyroclastic flows. The Neapolitans were merely glad to conclude that they had deserved to be saved because their saints had protected them and their sinners had repented. They did not know, and perhaps would not have wished to know, that violent eruptions of Vesuvius, and volcanoes like it, do not usually last more than a few days. Indeed, in spite of all the fervent religious input, the eruption in 1631 lasted just as long, if not rather longer, than usual.

Refugees and sinners

Many fugitives had made their own panic-stricken way out of the disaster area, but thousands of people injured or in shock had to wait for days to be rescued. The survivors were taken back to Naples to join the multitudes who had already sought a safe haven in the city. Many administrators were still try-ing to discourage these refugees from entering the city – now not primarily because they feared an epidemic, but mainly because they believed that a great influx of starving people would create a food shortage and public dis-turbances. However, the viceroy overruled these public servants and decreed that the refugees should be given shelter at his expense in mansions and var-ious public buildings. The authorities believed that sin was the root cause of the calamity, and they therefore engaged in good works. Individual citizens, and the religious orders offered as much food, shelter and solace as they could; and money was collected to provide for the poor, who had almost no means of sustenance. The administration paid special attention to young women made destitute by the crisis. "To preserve their modesty, over a hun-dred were placed in the safe keeping of the home for orphaned girls".

Special efforts were made to save prostitutes, whose sins had apparently contributed so much to the disaster. In and around Naples, 150 prostitutes were saved from their "evil life". Some were housed in a home founded especially to this effect by the Brothers of the Holy Rosary. At his own expense the viceroy himself gave shelter to 30 repentant prostitutes. Others were married off. "The cardinal", wrote Ceraso, "ordered six pious persons to go around the city . . . to solicit alms for these poor repentant women . . . [thus] a great number of young virgin girls and women of ill repute were saved from the hands of the Devil". Not to be outdone, the viceroy, in true administrative tradition, set up a committee to help them.

The churchmen widely publicized apparently miraculous survivals of churches and images. The stout construction of the buildings was almost certainly involved, but was not publicized. Where a church was saved, it was attributed to miracle; where a church was destroyed, the miracle could be detected when an intact image was discovered, such as, for instance, the statue of the Madonna in Santa Maria di Costantinopoli in Torre del Greco.

The varied effects of the pyroclastic flows could account for many of the miraculous survivals. The pyroclastic flows pass rapidly over any given spot and, therefore, any objects that ignite quickly can be easily damaged or incinerated, whereas other objects near by could remain intact. Stories of inexplicable survivals of both humans and objects soon gained currency. In a house in San Giorgio a Cremano, a pyroclastic flow did not even set fire to straw in the first room that it entered, although it burned everything in the second, where it seems to have stayed longer. In Torre Annunziata, a stoutly built mansion survived, but its main door was burned, and furniture and clothing inside were totally destroyed. A man galloped on horseback into Barra, where the townspeople discovered that he had been burned to death, whereas his horse had obviously survived. In Pietra Bianca, Abbé Braccini recounted that "a young woman with her little son at her breast fell to the flaming torrent, and her husband went to help her. He too fell down. Both were found dead, but the youngster was alive with the breast of his dead mother still in his mouth" (where her breast had protected the child). The survivors were yearning for miracles to compensate for so much mourning and so much devastation, and as a sign that the Lord had not totally abandoned them. They needed a future; thus, miracles were born.

Results of the eruption

After Christmas, several immediate tasks still faced the administration. All the measures had to be taken quickly and on an ad hoc basis, for there was no established administrative mechanism for undertaking these necessary tasks. The measures were probably successful, because there seem to have been few complaints or riots afterwards.

The viceroy organized squads of workmen and soldiers to clear the roads farther afield in eastern Campania, especially the Strada Regia, the main road from Naples to Salerno, along which most of the aid would have to pass. This road was opened as far as Torre del Greco on 27 December. The viceroy also sent engineer Aniello de Falco to assess the devastation all around Vesuvius. In fact, the immediate economic and social losses were incalculable and the administration had few real means of making any economic assessments that were more than impressions. De Falco estimated that losses in the area amounted to some 15–16 million ducats, plus about 1 million ducats a year lost in revenues from share-cropping, and Abbé Braccini later put the total figure at 25 million ducats. (The wealthiest landowners at the time had annual incomes of about 10000 ducats.)

The poor were even poorer than before, and many of the rich, who had previously owned much land and property, had been left with scarcely more than the title deeds to their possessions. Tales of once-rich landowners, wandering the streets of Naples begging for bread, clearly show where the sympathies of the establishment lay. It was rumoured, for instance, that the Prince of Ottaviano, who had enjoyed an income of 10000 ducats a year, had lost all his goods and chattels. The prince was inconsolable and was doing penance for his sins. So, too, was his household, either by choice or by force.

Many still feared that the eruption had expelled diseases that would infect the air and soon start to ravage the population. The reality was bad enough: famine threatened Naples during the week after Christmas. Thousands of refugees had to be fed, and bread in particular was in very short supply, because the flourmills around Vesuvius had been destroyed. On 27 December, the viceroy sent four galleys full of Neapolitan flour to be milled in Castellammare and Gragnano, which was then brought back to the city to be baked. Fortunately, this situation was soon eased when the flourmills at Torre Annunziata were repaired.

Commentators estimated that the eruption had killed between 4000 and 10000 people and a similar number of animals. These figures could only be guesses. No-one could be sure about the exact number of deaths because no-one had counted the population just before the eruption, and no-one could possibly have counted the corpses afterwards. However, it was certain that many towns were bereft of people and habitable houses, and that survivors were few indeed where the pyroclastic flows had devastated the populous towns around the volcano.

Most of the towns around Vesuvius had suffered great damage, and many of the ruins were buried deep in ash. The prime causes of the death and destruction were the pyroclastic flows that had emerged on the morning of

17 December. It was claimed that "there was mourning in every house". In Torre del Greco, for instance, Francesco Ceraso described how only the more substantial buildings had remained upright, although even these were often badly damaged. Among them were two churches, a couple of monasteries, the home of the Cardinal–Archbishop of Naples; and the palace of the Prince of Stigliano, the Lord of Torre del Greco, which stood on a hill overlooking the town. But the walls of all these buildings were as blackened as if they had been iron foundries.

Falling ash and pyroclastic flows had ruined the countryside and had smothered the fields across much of eastern Campania. Most of the beasts had been burned and those that had survived the eruption died of starvation when they could no longer reach the ash-mantled grass. The crops were dead. Fruit trees had been smashed in the orchards. Vast woodlands had been burned. Vines had been buried. Few farm buildings had stayed upright and farm implements had been destroyed. The farm products harvested during the previous autumn, and seed stored for the coming year, had been buried or burned. Huge tree trunks, uprooted by the pyroclastic flows, were floating in the sea and had brought inshore fishing to a halt. Several commentators soberly noted the great loss of wine. But, on the brighter side, some barrels of Greek wine had remained intact in a badly damaged house in Resina. Frat'Angelo wistfully recorded that "it had been sweetened and inflamed with colour, and it was not unpleasant to taste, although it had a slight smell of smoke". Perhaps it was almost a miracle?

Few commentators, even among the intellectuals, extended their search for the causes of the eruption beyond the established beliefs of the Counter Reformation. Giovanni Battista Manso was one of those who tried to account for different features of the eruption with some inkling of modern scientific methods, but, his meagre volcanological knowledge was derived from the classical texts. On the other hand, Manso might have discussed the great eruption with John Milton when he entertained the poet in Naples in 1638, and his account of events could possibly have provided Milton with some inspiration for his descriptions of Hell in *Paradise lost*.

Like all violent eruptions of Vesuvius, the most intense spasm of the outburst had calmed down by the end of the year, although it continued on a much reduced scale until it ended almost unnoticed during January 1632. The eruption had been catastrophic, but it had given off only $0.5\,km^3$ of ash and other incandescent materials, whereas about 30 times as much material had been expelled in AD 79. Nevertheless, Vesuvius had changed significantly.

Before 1631: the cone of Vesuvius that had developed before 1139

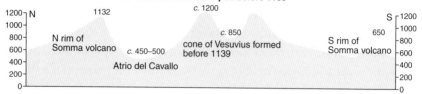

1632: the decapitated cone

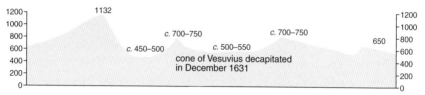

1752: renewed growth of the cone and lava accumulation in the Atrio del Cavallo

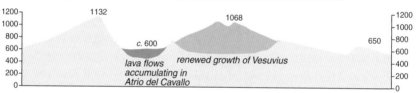

After 1944: more lava accumulation in the Atrio del Cavallo, increased growth of the cone, burial of the southern rim of the Somma caldera

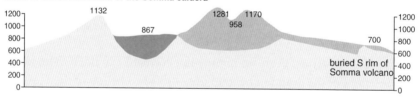

The main changes in the form of Vesuvius from before 1631 until 1944.

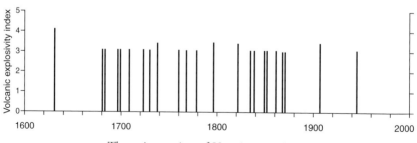

The main eruptions of Vesuvius since 1631.

The cone had lost some 450–500 m in height and its greatly enlarged crater was about 1650 m in diameter. The removal of so much material from the cone brought a mixed blessing to Naples. It was claimed that the warm morning Sun shone earlier into the narrow lanes of the old city and onto the shivering fishermen close inshore, although there were now more storms and wrecks in the bay because Vesuvius no longer protected it from the cold north winds. The volcanological result of the outburst in 1631 was that Vesuvius started a new style of activity, with more frequent eruptions, which helped the volcano to begin a distinguished career as a tourist attraction.

Future generations

The eruption of 1631 gave rise to an unexpected postscript that was quite at odds with the all-pervading beliefs that had gripped the whole district during the volcanic crisis. When the dust had settled, the viceroy, the Count of Monterrey, erected a series of inscriptions in Latin to commemorate the catastrophe and to warn the future citizens of what might lie in store for them. It is tempting to see the hand of Giovanni Battista Manso in these inscriptions. These commemorative tablets are remarkable documents for their time, which went much further than the truths received from the teachings of the Church. One of these tablets, in three marble panels, still stands in Portici at the corner of the Corso Garibaldi and the Via Emanuele Gianturco. In spite of the threat of accusations of heresy, the viceroy (or his administrators) made a logical connection between the natural warning signs of an eruption and the disastrous events that would ensue. The instructions given were clear. The viceroy urged the people to flee as soon as the mountain began to moan and groan and to shake the ground, because these were warning signs that a river of fire would suddenly appear, from which it would prove impossible to flee. Overleaf is a photograph of the inscription and a translation by Harry M. Hine (University of St Andrews, Scotland).

Further reading

Braccini 1631, 1632; Contreras 1990; Estrada 1982; Principe 1998; Rolandi et al. 1993; Rosi et al. 1993.

POSTERI POSTERI
VESTRA RES AGITVR
DIES FACEM PRÆFERT DIEI NVDIVS PERENDINO
ADVORTITE
VICIES AB SATV SOLIS IN FABVLATVR HISTORIA
ARSIT VESÆVVS
IMMANI SEMPER CLADE HÆSITANTIVM
NE POSTHAC INCERTOS OCCVPET MONEO
VTERVM GERIT MONS HIC
BITVMINE ALVMINE FERRO SVLPHVRE AVRO ARGENTO
NITRO AQVARVM FONTIBVS GRAVEM
SERIVS OCYVS IGNESCET PELAGOQ INFLVENTE PARIET
SED ANTE PARTVRIT
CONCVTITVR CONCVTITQ SOLVM
EVMIGAT CORVSCAT FLAMMIGERAT
QVATIT AEREM
HORRENDVM IMMVGIT BOAT TONAT ARCET FINIBVS ACCOLAS
EMICA DVM LICET
IAM IAM ENITITVR ERVMPIT MIXTVM IGNE LACVM EVOMIT
PRÆCIPITI RVIT ILLE LAPSV SERAMQ FVGAM PRÆVERTIT
SI CORRIPIT ACTVM EST PERIISTI
ANN·SAL·CIƆ IƆC XXXI XVI KAL IAN

PHILIPPO IV REGE
EMMANVELE FONSECA ET ZVNICA COMITE MONTIS REGII
PRO REGE
REPETITA SVPERIORVM TEMPORVM CALAMITATE SVBSIDIISQ CALAMITATIS
HVMANIVS QVO MVNIFICENTIVS
FORMIDATVS SERVAVIT SPRETVS OPPRESSIT INCAVTOS ET AVIDOS
QVIBVS LAR ET SVPPELLEX VITA POTIOR
TVM TV SI SAPIS AVDI CLAMANTEM LAPIDEM
SPERNE LAREM SPERNE SARCINVLAS MORA NVLLA FVGE
ANTONIO SVARES MESSIA MARCHIONE VICI
PRÆFECTO VIARVM

The commemorative Latin inscription in Portici warning future generations of the possible dangers from Vesuvius.

Future generations, future generations

This concerns you. One day carries a torch for another; and the day before yesterday for the day after tomorrow. Take note. Twenty times since the creation of the Sun, if history does not tell fables, Vesuvius has burned, always causing immense destruction to those who hesitate. I issue a warning, so that it may not catch people unawares henceforth. This mountain has a womb, pregnant with bitumen, alum, iron, sulphur, gold, silver, nitre, springs of water. Sooner or later it will catch fire, and as the sea flows in, it will give birth. But beforehand, it goes into labour, it is shaken, and it shakes the ground. It smokes, it flashes, it flames, it shakes the air. Horrendously it bellows, roars, thunders, keeps neighbours out of its territory. Flee while you may. It is on the point of bringing forth, it erupts, it spews out a lake mixed with fire. It rushes in a headlong flow, and prevents tardy escape. If it catches you, it is all over, you are lost. In the year of salvation 1631, 17 December, when Philip IV was king and Emanuele Fonseca and Zunica, Count of Monterrey, was viceroy, and when the disaster of earlier times was repeated, and when relief from the disaster was provided more humanely, because more generously.

If the mountain is feared, it saves; if it is scorned, it destroys the unwary and the greedy, for whom home and possessions are more important than life. You, if you are wise, must listen to the stone that shouts. Ignore your home, ignore your belongings, flee without delay.

Antonio Suares Messia, Marquis of Vico, prefect of roads

Chapter 8

The old cities rediscovered: antiquity protected

The ancient sites were discovered by accident and then pillaged by treasure hunters long before archaeologists could study them systematically.

It is hard to establish how many of the towns that had been buried in the ashes of the eruption of AD 79 were occupied again. The archaeological evidence is rather scanty, but it can be supplemented to some extent by the place names shown on surviving copies of Imperial Roman road maps. Between 1507 and 1547, a copy of one map of roads and places throughout the Roman Empire was published as the famous Peutinger Table by Konrad Peutinger in Augsburg in Germany. Herculaneum, Oplontis and Pompeii were marked on this map, which implied that these towns had been reoccupied after the catastrophe. But the evidence is not as strong as it might seem, because the Peutinger Table had a complicated and uncertain history. It was a thirteenth-century copy of a Roman route map that had been produced in the third or fourth century AD; and it is possible that this Roman map was itself a version of an even earlier work, which might have been drawn up before AD 79. Hence, the Peutinger Table itself does not provide wholly reliable evidence about the revival of the buried cities after AD 79.

Other evidence is often ambiguous. For instance, a century after the events, in *Ad se ipsum* (AD 166–178), the Emperor Marcus Aurelius became the first known author to describe Pompeii and Herculaneum as paradigms of death. On the other hand, there is some evidence that survivors did indeed return to the devastated areas and try to eke out a living above the ruins of these dead cities. Public baths and a cemetery existed on the site of Herculaneum in the third century; an industrial and residential complex had been constructed on the sites of Pompeii and nearby Scafati in the third and fourth centuries; and burial grounds dating from the third century have also been

Sketch of the exposure of the Temple of Isis at Pompeii in 1764, based on an illustration in Sir William Hamilton's Campi Phlegraei *(published in Naples in 1776).*

found in Pompeii and Herculaneum, as well as in Oplontis and Stabiae. Yet, the long period of anarchy associated with the fall of the Roman Empire in the West virtually ensured that the exact locations of the buried settlements were soon despatched to oblivion, their memory revived only from time to time by the briefest of references in literary works.

During the more peaceful periods of the High Middle Ages, resettlement began in the area on a larger scale. Resina grew up on top of the buried site of Herculaneum, Torre Annunziata was built over Oplontis, and Castellammare was built over Stabiae. The present town of Pompeii was constructed to the east of its old Roman site. The medieval villagers called this ancient site Civitas or La Città, without knowing its real name and significance.

Old stones come to light

From time to time throughout the Middle Ages, old decorated or inscribed stones turned up in the fields and were used to embellish new buildings. No-one understood their origin or cared about their importance. In the early sixteenth century, for instance, an old Latin inscription was noticed on a stone that had been used in the altar steps in the church of the Madonna in Scafati,

174

near Pompeii. It read: "Cuspius, son of Titus; Marcus Loreius son of Marcus". The inscription should have immediately suggested to the builders who had dug up the block that there might be ancient buildings in the vicinity. But no-one seems to have made any further enquiries.

More old stones were brought to light when drainage and irrigation channels were being cut to improve agricultural yields at the end of the sixteenth century. At La Città in 1592, the Roman architect, Domenico Fontana, was cutting a trench to make a water conduit when a few more old stones were dug up. Fontana did nothing about it. (In fact, the men had been digging within a few metres of the Roman amphitheatre in Pompeii.) In the last decade of the century, two more Latin inscriptions emerged, one dedicated to a Roman senator and the other to Jupiter. Then, in 1607, excavation of another trench revealed yet another old stone bearing the inscription *Decurio Pompeiis*. The name was now staring the diggers in the face. It revealed the site of the lost city of Pompeii, for the stone marked the site of the Pompeiian quarters of a *decurion*, a troop of ten cavalrymen. Again, nothing coherent was done about the discovery. All these finds were spread widely about the area, and they emerged irregularly and infrequently; the objects were not in themselves remarkable as works of art, and no precious metals came to light. Thus, there was no great "gold rush" to the sites.

The amphitheatre at Pompeii, with Vesuvius and the ridge of Monte Somma on the horizon.

This neglect was a great stroke of luck for the buried cities. Archaeology did not exist at that time, and all architects, from Michelangelo downwards, used old buildings merely as a cheap and easy source of good stones for their own masterpieces. Pompeii and Herculaneum could rest from spoliation for another century.

Excavations begin

The beginning of the excavations of the "lost cities" turned on a succession of chance events. During the War of the Spanish Succession from 1701 to 1713, Spain had to cede Campania to the Austrians. In 1709 one of these Austrians, the Prince of Elbeuf, had just begun to build himself a palace at Resina when his workmen unearthed several antique statues as they were digging a well. The prince immediately ordered the men to continue their excavations and then took charge of all the spoils. The finds included a statue of Hercules and also statues of three women, who were immediately called the Vestal Virgins. Prince Emmanuel presented these statues to his cousin, Prince Eugène, the famous general and ally of the Duke of Marlborough in the War of the Spanish Succession. When Prince Eugène died in 1736, the Vestal Virgins were sold to Frederick Augustus, the Elector of Saxony, whose young daughter, Maria Amalia, took a great fancy to them.

Meanwhile, in 1734, the Spaniards returned to power in Naples and Sicily, and Charles of Bourbon became king of what became known as the Kingdom of the Two Sicilies. He was to prove to be one of the ablest rulers that Naples ever had, although the competition for that honour was not intense. Charles had an artistic mother in Elizabeth Farnese, and early in 1738 he married an artistic wife, Maria Amalia of Saxony.

Charles of Bourbon at once started to modernize his capital and make it one of the finest in Europe. He widened the streets, took down the old city walls, built the opera house of San Carlo, and extended the royal palace in Naples. He also started work on other prestigious creations, such as the palace and royal-porcelain manufacture at Capodimonte; the palace at Caserta, in good hunting country that was to rival Versailles; a palace to house his Roman antiquities at Portici; and a huge hospital for the poor in Naples.

In 1738, the first official excavations at Herculaneum began, although their aim was not so much to unearth the ancient town as to dig up its treasures. Unfortunately, the town of Resina had been built on top of them, so

The Spanish Bourbon Kingdom of the Two Sicilies

In 1700, the Spanish Hapsburg dynasty ruling in Spain came to an end when Charles II died without an heir. With his death, too, the system of ruling Naples and Sicily through Spanish viceroys also ceased. The rival monarchs of Europe vied with each other to grasp the chance to expand their power in Spain and its vast dominions. Of course, the people living in those countries were not consulted. Louis XIV, the Bourbon King of France, designated his grandson, Philip, as the new King of Spain. The Austrians and the British did not approve of this arrangement and fought the War of the Spanish Succession to try to prevent it. After great battles and tortuous diplomatic changes, Louis XIV's Bourbon grandson did become Philip V of Spain, and Naples fell into Austrian hands for several decades.

The dynastic musical chairs continued. Naples, southern Italy and Sicily raised the status of a kingdom called the Two Sicilies and, in 1734, Charles of Bourbon, the third son of Philip V, was made king. Charles's two elder brothers, Louis and Ferdinand, both succeeded Philip V as King of Spain, but neither produced an heir. Thus, in 1759, Charles of Bourbon succeeded to the Spanish throne and he had to leave Naples for Madrid. The experiences of the previous 60 years dictated that the two thrones should be kept separate in future. A monarch therefore had to be found for the Two Sicilies who was unlikely to inherit another kingdom.

However, the Spanish Bourbons had been plagued by insanity. Philip V and his son, Ferdinand VI of Spain, had been seriously afflicted. Charles's eldest son, Philip Anthony, was also insane and had already been declared unfit to rule. Charles took his second son, Charles, with him to Madrid as the heir to the Spanish throne. This left his eight-year-old third son, Ferdinand, to succeed to the throne of the Two Sicilies. Haunted by the prospect that young Ferdinand might also become tainted with insanity, his advisers prescribed an incessant round of outdoor activity for him. It suited him well: he spent virtually his whole life hunting, shooting or butchering his prey. His indoor activities seldom extended beyond boisterous games and siring children. He could never bear to be alone, not even when enthroned upon the toilet, as his brother-in-law, the Austrian Emperor Joseph II, discovered to his own cost in 1769. Perhaps it was this unhappy experience that made the emperor comment that Ferdinand was "quite ignorant of the past and present, and has never thought about the future". This dolt was to reign until 1825, although he had been forced to take refuge in the Sicilian part of his kingdom between 1806 and 1815, when the French had occupied Naples. Nevertheless, for a monarch of the period, Ferdinand had the extraordinary saving grace of being very kind hearted; and he was affectionately known throughout the kingdom, as Nasone ("Big Nose"). In 1768, he married Maria Carolina, a daughter of the Austrian Empress Maria Theresa. Maria Carolina took little more than a decade to take charge of Naples and free her husband for yet more hunting. She was to become a great friend of Horatio Nelson and Sir William Hamilton's second wife, Emma, while the refined, artistic and intelligent Sir William was to be one of Ferdinand's closest companions for 36 years.

that shafts, galleries and long tunnels had to be dug to reach the spoils. The Roman treasures that were unearthed gave the fledgling kingdom further prestige by establishing its links with the glorious Roman past.

In 1748, royal workmen unearthed some Roman columns and paintings at the site of ancient Pompeii. The king ordered excavations to begin there at once. Pompeii proved to be a much easier prey than Herculaneum, for it was set in open fields and covered with easily removable ash and pumice. However, to direct the excavations, King Charles brought over from Spain the engineer Alcubierre, who matched inexperience with so much incompetence that it is amazing that any relics ever managed to emerge intact. Thus, the buried cities were discovered too early for modern archaeological methods of excavation to be applied to their exploration from the outset. For decades, the precious sites were ransacked in unscientific treasure hunts that probably caused as much archaeological damage in the Roman towns as the great eruption itself had caused to their fabric.

Charles decreed that the treasures of Pompeii should be removed and brought with those of Herculaneum to the royal palace and museum constructed specially for the purpose at Portici. During the next few decades, the excavations switched between the two ancient towns. In 1750, Herculaneum yielded one of its greatest treasures: the Villa dei Papiri. This was a great library containing thousands of papyrus scrolls that had been charred by the pyroclastic flows in AD 79. In 1750, they looked like burnt sausage rolls that crumbled when anyone tried to unwrap them. The excavators mostly left them alone, which was fortunate for posterity because they can now be unravelled and analyzed by sophisticated modern techniques and they promise to lay bare some of the most important literary legacies of the ancient world that have ever been brought to light.

Excavations at Pompeii

The excavations at Pompeii revealed more obviously spectacular finds in old streets that were lined with villas full of paintings, statues and mosaics, including the Villa Felix as only the first of many to be revealed. The renown of the sites began to spread among the European intelligentsia. In 1755, the first volume of the antiquities of Herculaneum was published, and in the same year Robert Adam, who was to bring so much of the classical spirit to Britain, was given a quick tour of the site. Even more important in the short term

The Vesuvius gate at Pompeii, to the left of which lie some of the very thick layers of pumice and ash that had to be removed before the buildings were revealed. Vesuvius forms the horizon on the right.

was the visit of the eminent German art historian, Johann Joachim Winckelmann. He was appalled by the slovenly and anarchic methods of excavation employed, and he said so in two open letters published in 1762. Although his protests earned him the disapproval of the Neapolitan Court, they publicized the extraordinary historic finds throughout Europe, and even induced the excavators to improve their methods.

After young Ferdinand was designated King of Naples in 1759, the regent Tanucci continued the excavations. The publication of their early results had an enormous influence on European fashion and taste. In 1763, Tanucci ordered that the excavations should be concentrated on Pompeii, where the treasures could be unearthed more easily. The gems of Pompeii soon began to emerge. In 1764, the Odeon and the Temple of Isis were revealed; in 1767 it was the turn of the barracks of the gladiators; and in 1771, the Diomedes Villa was exposed. However, although the excavations became rather more methodical, anarchic digging, pilfering, looting and wanton destruction predominated. Walls were knocked down, and statues and jewellery were stolen and sold off to anyone who fancied a piece of Roman history. For instance, when Goethe went to Herculaneum on 18 March 1787, he complained that it was "a thousand pities that the site was not being excavated methodically by German miners, instead of casually being ransacked, as if by brigands, because many noble works of antiquity must have been thereby lost or ruined". Goethe also recognized the extraordinary importance of the buried cities. "There have been", he wrote, "many disasters in this world, but few that have given so much delight to posterity, and I have seldom seen anything so interesting".

The revelations from Pompeii and Herculaneum helped to popularize the art of the ancient world; and classical art flourished and reached the height of fashion for the first time in 1250 years. After William Hamilton's appointment as British envoy in 1764, he became one of those who did most to spread knowledge of the extraordinary revelations from the ancient towns, and he diligently educated the increasing number of visitors to his Neapolitan residence. Hamilton also bought up vases and pieces of ancient statuary, and sold them to collectors on the Grand Tour and even to the British Museum.

The excavations received a further spur during the Napoleonic occupation of Naples from 1806 to 1815, when the new French King of Naples and Sicily (Joachim Murat) and his wife Caroline Bonaparte actually paid for the work with their own money, and took an active interest in the results. The excavations continued when Ferdinand returned from his Sicilian exile in 1815. His successor Francesco I followed the work closely and he was responsible for the reopening of the diggings at Herculaneum, which had been more or less abandoned since 1763. Each new revelation stimulated further excavations, and none more so than the discovery of the superb House of the Faun in Pompeii in 1830.

Charles Dickens saw Pompeii in a rather different light. On his visit in February 1845, he was naturally most fascinated by the ordinary aspects of life revealed in Pompeii:

> Stand at the bottom of the great marketplace [Forum] of Pompeii, and look up the silent streets through the ruined temples of Jupiter and Isis, over the broken houses with their inmost sanctuaries open to the day, away to Mount Vesuvius, bright and snowy in the peaceful distance; and lose all count of time, and heed of other things, in the strange and melancholy sensation of seeing the Destroyed and the Destroyer making this quiet picture in the Sun. Then . . . see, at every turn, the little familiar tokens of human habitation and everyday pursuits; the chafing of the bucket rope in the stone rim of the exhausted well; the track of carriage wheels in the pavement of the street; the marks of drinking vessels on the stone counter of the wine shop; the amphorae in the private cellars, stored away so many hundred years ago, and undisturbed to this hour.

The role of Giuseppe Fiorelli

The unification of Italy in 1861, and the development of real archaeological methods, gave a further stimulus to excavations at both sites. Giuseppe Fiorelli was appointed director of excavations, which he undertook with scientific rigour and the help of 500 workmen. It was he, in February 1863, who developed the method of taking plaster of Paris moulds of the people found in their death throes in Pompeii, which are surely among the most vivid and poignant archaeological relics in the world. Fiorelli realized that the dry and rigid ash and pumice had enshrouded and moulded the bodies of the victims where they had fallen. Subsequently, the softer parts of their bodies had decayed, and had therefore left behind a void containing only the skeletons of the victims. But the ash and pumice had remained firm enough to retain the moulds of the original shapes of the bodies and their clothing. If the excavators were careful, therefore, these original shapes could be restored and revealed. Fiorelli hit upon the brilliant idea of adapting the method used for casting bronze statues. Whenever Fiorelli's excavators discovered a hole in the ash and pumice, they carefully poured liquid plaster of Paris into the void and waited for it to set. Then, equally carefully, they scraped the cover of ash and pumice away from the solidified plaster. They then had detailed models of the bodies of the original victims in the very moment when they died. The suffering of the victims is stamped on their every gesture as they tried to protect their mouths from the hot ash, craned their necks to breathe, or strained every muscle in their dying agony.

Fiorelli used this method for the first time on the remains of four victims found in what was to be called in their honour the Vicolo degli Scheletri ("alley of the skeletons"). The man leading the little group in the alley had been thrown down by the pyroclastic flow and he died trying to raise himself onto his elbows and to protect his head with his cloak. His two daughters died a few paces behind him, and the cast showed that the younger girl had her hair in plaits. His wife had raised her skirts up as she ran and had been thrown down behind her daughters. She was carrying the family's most precious possessions in a small bag – a little amber statuette, a small silver plate and a silver mirror. Thereafter, rewarded by such encouraging detail, the method pioneered by Fiorelli became standard practice for over a century. It has recently been modified so that a clear resin is used instead of plaster of Paris, which enables, for instance, jewellery and skeletal bones to be seen.

The human remains in Pompeii are much better preserved than those in

Herculaneum overlain by pyroclastic-flow deposits, with the modern town of Ercolano (Resina) constructed upon them.

Some of the deposits that buried the Villa of Poppaea Sabina at Oplontis. Coarse airborne ash and pumice are exposed in the lower third of the section, and finer fragments from the pyroclastic flows occupy the upper two-thirds of the section.

Herculaneum, and the Fiorelli technique could be used only rarely here. The corpses were buried and often crushed under the thick layers laid down by the pyroclastic flows. More importantly, Herculaneum has subsided by about 4 m since the eruption, carrying the bodies of the victims and the enclosing volcanic deposits beneath the water table. As a result the flesh and clothing of the victims have rotted away so that only their skeletons have survived.

Herculaneum also presents even greater problems for excavators, because Resina has been built directly over the Roman site. Excavations there are still difficult and costly, and often have to be done along tunnels under the new town lying above the site. However, some 85 m along the old shore of the town has now been laid bare and the skeletons of many of those killed by the pyroclastic flow have been revealed.

There are problems at Pompeii too. More tourists now often visit the ruins in a day than ever thronged the streets of the ancient town. Pompeii was bombed during World War II, and it was also damaged by an earthquake in 1980. The sites have suffered from vandalism and theft ever since they were first brought to light, and there is little evidence that such crime is declining.

Many houses have to be locked for their own protection, and the custodians cannot always be bothered to open them for interested visitors. The result is that Pompeii has already lost the charm and Roman atmosphere that still manage to prevail in Herculaneum and in the exceptional Villa of Poppaea Sabina at Oplontis, which has been perhaps the greatest revelation of recent decades. It is also worth stressing that, although scientific excavations have been going on for over a hundred years, a quarter of Pompeii and two thirds of Herculaneum have yet to be revealed.

Further reading

Bowersock 1978; Sigurdsson 1985.

Chapter 9

Hamilton and Vesuvius:
volcano-watching

The British envoy taught himself to study volcanoes by watching the behaviour of Vesuvius. His letters to the Royal Society in London not only gave Vesuvius international publicity but also laid some of the foundations of volcanology.

The eruption of 1631 cleared the main vent of Vesuvius so effectively that it could erupt far more easily than it had done for centuries. The volcano then embarked upon a period of over 300 years when the magma was lying not far below ground and could easily surge up the cleared vent and erupt onto the surface at irregular frequent intervals. This new behavioural pattern of almost continual activity proved to be an enormous boost to the nascent study of volcanoes and a great boon to the Neapolitan tourist trade. The repeated agitation gave the Campanians and their visitors continual thrills to which they rapidly grew accustomed, if not actually addicted. Vesuvius became not only part of the scenery but also an active part of their lives. Thus it was that Vesuvius launched first Francesco Serao, in 1737, and then William Hamilton, from 1764, on studies that have become classics of volcanology. It was Vesuvius and the Campanians that made Naples one of the most glamorous destinations on the aristocratic Grand Tour.

Questions of pedigree

William Hamilton was one of the most intriguing British personalities of the eighteenth century: a cultured and intelligent aristocrat, one of the founders of volcanology, an eminent collector of antiquities, a constant companion of a royal buffoon, and the notorious third leg of a *ménage à trois* with his second wife, Emma, and Admiral Horatio Nelson.

185

AS

Sketch of Vesuvius, based on an engraving in Bottoni (1692). The ridge of Monte Somma, forming the higher northern parts of the rim of the old Somma caldera, rises on the far left. The lower rim of the caldera is almost buried by the fragments erupted by Vesuvius, but it is still visible alongside the Hermitage. Eruptions have started to build up the new cone of Vesuvius within the crater of the old cone of Vesuvius, which was decapitated in 1631. Three vents are shown in activity.

William Hamilton was born near Henley on 13 December 1730. He was the third son of Lord Archibald Hamilton, the seventh son of the third Duke of Hamilton. Lord Archibald was a placid nonentity, but his wife Lady Jane was the mistress of Frederick, the Prince of Wales. The infant William was the playmate of Frederick's oldest son, Prince George, who was to succeed to the British crown in 1760 as King George III.

Like many a younger son, William had little personal income when he left Westminster School, and he had no choice but to join the Church or the armed services. Thus, he spent ten years in the Foot Guards, where he became an excellent shot, although he always had more enthusiasm for his violin lessons and artistic company than for military exercises. William inherited little from his parents and had to find a new career as soon as he left the army in 1757. Since work in trade was inconceivable for an aristocrat, he also had to find a rich wife. In 1758, he duly married Catherine Barlow, heiress to valuable estates in Pembrokeshire, Wales. In 1761, in accordance with the

Eruptions of Vesuvius after 1631

The eruption in 1631 brought about a distinct change in the behaviour of the volcano and initiated a new phase of almost continual persistent activity, often with an open vent, which went on until 1944. It may or may not be a coincidence that this period lasted almost as long as that between 787 and 1139. Between 1631 and 1944, Vesuvius expelled a total of some 5 km^3 of volcanic materials and the longest dormant periods rarely lasted a decade. The volcano gave off large lava flows and exploded much ash, pumice and cinders. During this period, Vesuvius displayed one of three types of eruptions, which have been termed "persistent", "intermediate" and "final".

Persistent activity   When activity resumed after a dormant period, lava issued either from the crater or from fissures on the sides of the cone itself, and explosions of ash and cinders then quickly followed. These eruptions commonly had a volcanic explosivity index of 1–3. Such effusive–explosive Strombolian activity often continued for years at a time and lasted for more than half the period from 1631 to 1944. These were the constructive periods when Vesuvius was growing in height and volume. Thus, the long period of Strombolian activity between 1875 and 1906 built up Vesuvius until, in May 1905, it reached 1335 m, the maximum height recorded in modern times.

Intermediate eruptions   The persistent moderate activity was punctuated by shorter stronger outbursts, usually lasting no more than two or three weeks, that have been called "intermediate eruptions". Since 1631, 44 of these "intermediate eruptions", with a volcanic explosivity index of 2 or 3, have been identified. They often produced greater volumes of lava, including huge lava fountains widespread clouds of ash that showered down on the neighbouring towns and villages. They were always followed either by a return to persistent moderate activity or by a sudden change of gear that led to even more powerful eruptions.

Final eruptions   The more violent "final eruptions" were more vigorous, much shorter and less frequent than any other form of activity on Vesuvius since 1631. They are so named because they always brought the whole phase of volcanic hostilities to a resounding close. Since 1631, 22 final eruptions, with a volcanic explosivity index of 3 or 4, have been identified. They not only expelled copious lava flows but also developed towering plumes of ash and cinders that rose as much as 15 km above the volcano and typically ended with a large explosion of white ash. These eruptions usually lasted less than two weeks, but they often caused the most widespread damage, and sometimes loss of life. At the end of these final eruptions, Vesuvius would embark upon another dormant period and the whole process would start all over again. The latest final eruption occurred in 1944, and the period of repose since that date has been by far the longest since 1631. Thus, Vesuvius could have entered another prolonged dormant period similar to that which occurred after 1139.

Engraving of Sir William Hamilton by Hugh Douglas Hamilton
(© Crown copyright: UK Government Art Collection).

democratic principles of the day, he was given the safe parliamentary seat of Midhurst in Sussex, but he took little interest in the duties required. More-over, Catherine had delicate lungs and it soon became clear that she could not tolerate the damp British climate. William therefore obtained the post of British Envoy Extraordinary to the Kingdom of the Two Sicilies. The Hamiltons arrived in the capital, Naples, on 17 November 1764.

The envoy in Naples

In the mid-eighteenth century, Naples was a busy, cultured and thriving city with a population that was reputed to be about 350 000. There were many nuns, monks, priests and soldiers, and some 40 000 boisterous *lazzaroni* (the Neapolitan poor), living rough and entirely dependent upon their considerable ingenuity. Most visitors thought that Naples was "a paradise inhabited by devils" living only for the present, selling anything that they could lay hands on – and, above all, wanting to enjoy every opportunity that life offered. Such behaviour astonished visitors from the north, who, even before the Industrial Revolution, expected the working classes to be more subdued. However, in spite of Charles of Bourbon's efforts, Naples was a still political backwater without the glamour attached to the more important capitals, such as Paris, Vienna and Saint Petersburg, to which Hamilton sought a nomination in vain. Thus, he remained in Naples for the next 36 years.

Britain was the main trading partner of the Kingdom of the Two Sicilies, and the British envoy had to act as a conduit of information. He had to send frequent reports to London, for which he received an expenses allowance of about £500 per annum, plus a salary of £5 a day. Although these payments were more than a farmer on his Welsh estates would earn in a decade, they

Sketch of Vesuvius based on an illustration published in Serao (1738). A century of eruptions of Vesuvius have rebuilt the cone decapitated in 1631, and lavas are accumulating in the floor of the Atrio del Cavallo, although the lower southern rim of the old Somma caldera is still visible. On the lower right, the monastery of the Camaldoli della Torre forms a prominent landmark on an old lateral cone erupted on Somma volcano that lava flows have never swamped.

had to be set against the unavoidable expense of entertaining every passing British nobleman on the Grand Tour.

Hamilton established his official Neapolitan residence in the Palazzo Sessa, where he soon began to organize frequent musical evenings. It was there, on 18 May 1770, that he enjoyed a privilege given to few envoys of any government, and almost certainly never granted to a volcanologist. He and the musician, Pugnani, played the violin, and they were accompanied on the harpsichord by Leopold and Wolfgang Amadeus Mozart.

Within a few months of reaching Naples, Hamilton began to pursue the three main themes of his Neapolitan life: shooting, vases and Vesuvius. Being an excellent shot and horseman, he soon became the favourite and influential hunting companion of young King Ferdinand; he started his collection of antique vases, many of which he sold to the British Museum, where they formed one of the foundations of its collection; and he began his lifelong fascination with Vesuvius.

Naples was the only European capital where a passion for volcanic eruptions could become an obsession. The most famous volcano in Europe was on its very doorstep, and Vesuvius was frequently turning on displays of a beauty that had not been surpassed for centuries. Hamilton began his daily observations of the volcano as early as June 1765. He had no training; there was not much training to be had, for the Earth sciences still hardly existed, and had certainly not established a methodology. He soon took the Villa Angelica, near Portici, where he installed his observatory to watch the volcano more closely, and he often climbed Vesuvius, trying to describe and make some sense of what he saw. He then wrote graphic letters to the Royal Society in London, which became some of the classic papers of volcanology and made him one of the world's foremost experts on volcanoes.

Trespassing on Vesuvian territory

The Hamiltons arrived in Naples during the early stages of a marked expansion of population into the Campanian countryside. At the beginning of the eighteenth century the total population of the whole Vesuvian area probably did not exceed 25 000 and the volcano rose out of a mainly rural landscape of small villages surrounded by vineyards, orchards and market gardens, which had at last recovered from the devastation caused by the great eruption of 1631. Then, within a few decades, the population of the area increased so much that many villages grew into bustling towns, and a rash of buildings began to approach the dangerous volcanic domain. The increased frequency of the eruptions in no way deterred the new settlers. Even when eruptions destroyed settlements such as Torre del Greco, the citizens would not budge: they usually just built new homes on top of the lava. It was as if the inhabitants were throwing down the gauntlet and challenging Vesuvius to do its worst. In fact, given the moderate nature of most of the eruptions, the long-term benefits to the region probably outweighed the losses. The chief benefits came to the agricultural lands, where the ash soon weathered and actually increased in fertility within a few years. On the other hand, the lava flows destroyed towns, buildings and crops, but the areas devastated were limited to narrow zones, so that the chances of any given spot being overwhelmed during a human lifetime were relatively small. Moreover, neither falling ash nor lavas usually caused much loss of life, and only then in a few villages at a time. Between the great eruption in 1631 and the latest eruption in 1944, Vesuvius claimed fewer than 300 victims. Overall, therefore, most of the inhabitants who accepted the challenge of Vesuvius were quite successful. Nowadays, the proof is there for all to see, for this area contains some of the most densely populated districts in all of Europe.

The expansion towards Vesuvius began in 1709, when the Prince of Elbeuf started to build his palace at Resina. This initiative launched two fashions among the wealthy Campanians. The prince's workmen uncovered Roman remains, which led to the excavation of the towns buried in AD 79, and it also led to the craze for building fine villas in the attractive countryside on the slopes of Vesuvius. The local aristocracy were following a trend that was developing throughout Europe for building ostentatious country homes as the landed classes started to spend their increasing income from their great estates. In Campania, these fashions flourished even more after the Spanish Bourbons replaced the Austrians in 1734. In 1738, Charles of Bourbon himself boosted the trend when he built the royal palace at Portici, and he restricted religious constructions and actively encouraged secular building projects by offering planning permissions and even tax concessions. Some villas, especially those near the Bay of Naples, were built simply as pleasant recreational country houses, some as property investments, and others (notably those inland) as hunting lodges or agricultural or market-gardening centres – although many combined several of these functions. Eventually, about 150 villas were built around Vesuvius. There was a concours d'élégance, too, as fine architects such as Sanfelice at Portici, Fuga at Favorita in Resina and the Vanvitelli brothers at Campolieto, constructed gems in the elegant Rococo style and set them amid well planned vistas. By the end of the eighteenth century, Portici had 38 villas, San Giorgio a Cremano 35, Resina 23, and Torre del Greco 20. No-one seemed to be worried that most of the villas constituted large and immobile capital investments, many of them on the southern slopes of Vesuvius, which were most vulnerable to volcanic attack.

On the other hand, many villas were situated within reach of the Strada Regia ("royal highway"), extending from Naples to Calabria and thence to Sicily, which was the chief mainland artery linking the two sections of the kingdom. The villas attracted a whole gamut of retainers – servants, suppliers, workmen, farmers, gardeners – who settled in the nearby towns, which were already expanding, as economic conditions improved in the more stable social, political, and more salubrious, environment. For all their agricultural produce, they had in nearby Naples a vast market. Before the end of the eighteenth century, these increasingly rich settlements along the artery came to form what was widely known as the Miglio d'Oro ("golden mile").

Thus began a cumulative effect of population expansion that has continued to the present day: the greater the number of inhabitants, the greater the demand for produce and services, and the more people were attracted to the growing settlements. In 1778, perhaps almost 40 000 people were clustered in these coastal towns; and the population had already almost doubled by the time the first railway in Italy was opened in 1839 from Naples to Portici. This

The Villa Favorita and the old Strada Regia at Ercolano (Resina).

was in spite of the fact that, in the previous 80 years, Vesuvius had unleashed six sizeable eruptions (and many more of lesser vigour), which had destroyed part of Torre del Greco and threatened some of its neighbours.

Now isolated monuments to a bygone age, the villas have lost their former glory. Fewer than a score are still in private hands, some have been abandoned and the majority have been divided into multiple dwellings; their parks have provided tempting sites that construction companies could hardly resist. The building stones of Neapolitan Yellow Tuff and their fragile cover of stucco have long since begun to decay. Only a few fortunate villas have been adapted to functions that could preserve them: for example, the Villa Campolieto is now an art gallery and exhibition centre, and the royal palace at Portici has housed part of the University of Naples since 1873.

However, the increase in population that the villas stimulated has continued apace. Over 600 000 people – many more than ever before – now live around Vesuvius, and settlements have been proliferating up to the volcano and lapping around the main cone as if it will never erupt again. Such prospects were inconceivable when Hamilton witnessed his first eruption of Vesuvius.

The eruption of 1766

In September 1765, William accompanied the French astronomer, Jérôme Lalande, to the summit of Vesuvius, and noticed that the crater was giving off more smoke than usual. In October, "occasional puffs of black smoke shot up to a considerable height in the midst of the white, which symptom of an approaching eruption grew more frequent daily". In early November, Hamilton climbed the snow-covered mountain and saw that, since his previous visit, "a hillock of sulphur had been thrown up . . . and a light blue flame issued constantly from its top". As he was examining this hillock, a loud explosion sent a shower of stones from its crater and made him "retire with some precipitation" and taught him to be more cautious in future.

Activity increased throughout the winter, and ash and fumes caused much damage to the vineyards around the volcano. In March 1766, the black smoke took the form of a pine tree, and the loud explosions of glowing ash created a splendid firework display every night. Nevertheless, Hamilton was soon up the mountain once again, this time acting as the guide for Lord Hillsborough, whose wife had just died. William thought that the trip would

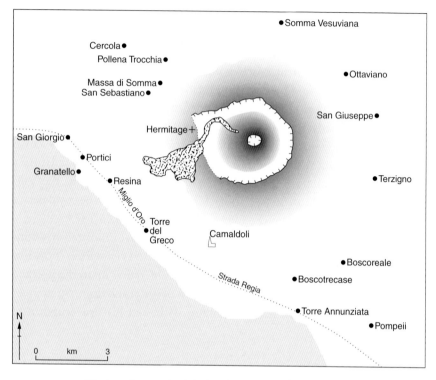

The lava flows emitted during the eruption of 1766–1767.

help distract him from his grief. In fact, they were lucky to escape injury, for the volcano erupted vigorously the next day.

On Good Friday, 28 March 1766, Hamilton was delighted when a strong explosion hurled masses of red-hot cinders into the air, and a small earth-quake announced the arrival of the first lava flows at 7 p.m. He and a party of Britons immediately left Naples to satisfy their curiosity. They spent the night on the volcano and witnessed some moderate emissions of lava and ash that marked the beginning of one of the "intermediate" eruptions of the volcano:

> The lava ran near a mile [1.6 km] in an hour's time . . . I approached the mouth of the volcano, as near as I could with prudence. The lava had the appearance of a river of red-hot and liquid metal, such as we see in the glass[-blowing] houses, on which were large floating cinders, half lighted, and rolling one over another with great precipitation down the side of the mountain, forming a most beautiful and uncommon cascade.

194

On 30 March, the volcano threw splays of red-hot stones, like an enormous firework. Hamilton passed the night of 31 March on the volcano, and saw perfectly transparent red-hot stones, some perhaps weighing a tonne, being hurled 60 m into the air.

On 10 April, lava stopped erupting from the flank facing Naples, but broke out more vigorously from a long fissure, 800 m from the main crater, on the southeastern slopes facing Torre Annunziata. As the eruptions of ash and pumice diminished, the lava flow became the main centre of attraction. On 12 April, Hamilton spent a day and night alongside the flow and climbed up to its source, which was a vent situated some 800 m below the main crater. "It was like a torrent, attended with violent explosions, which threw up inflamed matter to a considerable height, the ground quivering like the timbers of a watermill." The flow was so hot that he could not approach within 3 m of it and, although it looked as fluid as water, it was stronger than it seemed. He prodded the flow with a long stick and threw large stones onto it, but they made little impression on the surface and floated away out of sight. The flow "ran with amazing velocity; I am sure the first mile [was covered] with a rapidity equal to that of the River Severn, at the passage near Bristol". Although the flow emerged as a thin stream, it soon spread out into a sheet, 6 km long and nearly 3 km wide. After the lava had flowed for about 100 m, it "began to collect cinders, stones, etc.; and a scum was formed on its surface, which in the daytime had the appearance of the River Thames, as I have seen it after a hard frost and a great fall of snow, when beginning to thaw, carrying down vast masses of snow and ice".

The lava rushed down hill unabated towards the cultivated slopes of the mountain. On the night of 12 April:

I saw it . . . unmercifully destroy a poor man's vineyard, and [it] furrowed his cottage, notwithstanding the opposition of many images of Saint Januarius [San Gennaro] that were placed upon the cottage and tied to almost every vine. [But, at its snout] the lava was no longer liquid, but resembled a heap of red-hot coals, forming a wall in some places ten or twelve feet high, which rolling from the top soon formed another wall . . . advancing slowly, not more than about 30 feet an hour.

Then the vigour of the eruption began to decline. By 3 June, the flows were moving only slowly, and in November they stopped altogether, although the crater continued to give out fumes until 10 December 1766.

These extraordinary sights induced Hamilton to write an account to the Royal Society in London. It earned an enthralled reception and he was made a fellow of the Royal Society in September 1767. He could have received no greater accolade or encouragement to pursue his research. That very same year, King George III made him Envoy Extraordinary and Plenipotentiary, which gave him added prestige and boosted his salary by £3 a day.

Neither Vesuvius nor the envoy extraordinary rested on their laurels for long. Indeed, the volcano even issued a warning to young King Ferdinand. An emission of an invisible noxious gas, broke into the chapel in the palace at Portici. An intrigued servant opened the door and was immediately overcome by the fumes. In a paddock near by, His Majesty saw one of his dogs drop down; and a lad who went to pick it up lost consciousness as well. Luckily, another servant realized what was afoot and he managed to drag the lad and the dog away from the emission. His Majesty was so impressed that he even remembered to tell Hamilton.

The eruption of October 1767

During the spring of 1767, the volcano seemed to be preparing for a greater effort. The explosions of ash, cinders and pumice in the main crater built up a little cone that soon grew higher than its rim. On 7 August, lava emerged from this cone, filled the main crater of Vesuvius and, on 12 September, spilled over its brim down the side of Vesuvius. Frequent explosions from the cone threw out volumes of red-hot cinders, which accumulated so quickly that it reached a height of 60 m by 15 October. Hamilton sketched the growth of this cone from the Villa Angelica at Portici. He soon became convinced "that the whole of Mount Vesuvius has been formed in the same manner", and sent his drawings to the Royal Society to demonstrate it. This was, in fact, a more important observation than Hamilton at first realized. It gave the lie to a belief commonly held at the time that volcanoes were formed because the land first swelled up like a boil, before fragments escaped from hole in the summit. The sketches effectively demonstrated that eruptions themselves built up volcanoes by accumulating layer upon layer of fragments above and around their vents.

Eventually, at 7 a.m. on Monday, 19 October 1767, Vesuvius began a "final eruption" that was its most violent outburst for more than a hundred

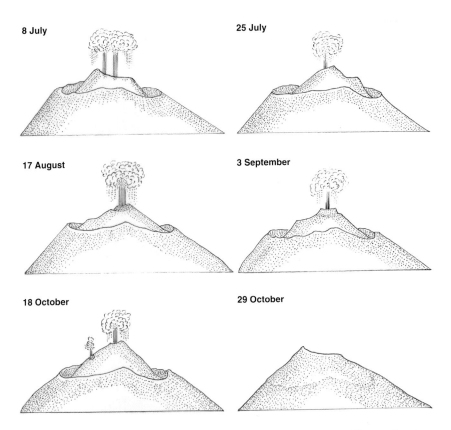

Sketches of successive profiles of Vesuvius during the eruption in 1767. They are based on plate III in Hamilton (1768). They illustrate the changes in the form of the summit area. The diagrams seem to have been drawn from careful free-hand sketches that are not all on the same scale. Thus, for example, the summit on 29 October 1767 was some 65 m higher than it had been on 8 July.

years. Hamilton's adventures and observations gave him the raw material for another letter to the Royal Society of London:

> From the top of the little mountain issued a thick black smoak, so thick that it seemed to have difficulty in forcing its way out; cloud after cloud mounted with a hasty spiral motion, and every minute a volley of great stones were shot up to an immense height in the midst of these clouds; by degrees, the smoak took the exact shape of a huge pine tree, such as Pliny the Younger described in his letter to Tacitus.

As the wind carried the crest of this cloud out westwards over Capri, "I

AS

Sketch of the eruption of Vesuvius in 1767 seen from Naples, based on an illustration to Hamilton's letter to the Royal Society of London (Hamilton 1768). Lava fountains spurt vertically out from the crater of Vesuvius, while a lava flow erupts from a vent near the base of the cone. The Hermitage is protected by the southwestern rim of the Somma caldera, although the lava spills out over the lower southern rim and threatens the coastal towns. The monastery of the Camaldoli della Torre rises on the right on an old lateral cone that erupted on Somma volcano long before the present cone of Vesuvius was formed. The Royal Palace of Portici lies just below it and to the left.

warned my family not to be alarmed, as I expected there would be an earthquake at the moment of the lava's bursting out". Here Hamilton was misinformed, for the lava issued forth at 8 a.m with neither noise nor quaking, while the explosions from the crater became much less violent. Believing (wrongly) that the danger had been reduced when the lava had erupted, Hamilton "went up the mountain immediately, accompanied by one peasant only", probably his usual guide, Bartolomeo Pumo ("Tolo"). They reached the Atrio del Cavallo, the valley between Monte Somma and Vesuvius, where Hamilton went on studying the lava.

What is it like for an aristocrat to be caught up in an eruption of lava?

On a sudden, about noon, I heard a violent noise within the mountain, and . . . about a quarter of a mile off the place where I stood, the mountain split, and, with much noise, from this new mouth, a fountain of liquid fire shot up many feet high, and then, like a torrent, rolled on directly towards us. The earth shook, at the same time that a volley of

198

pumice stones fell thick upon us; in an instant, clouds of black smoak and ashes caused almost total darkness; the explosions from the top of the mountain were much louder than any thunder I ever heard, and the smell of sulphur was very offensive.

My guide, alarmed, took to his heels; and I must confess that I was not my ease. I followed close, and we ran near three miles without stopping; as the earth continued to shake under our feet, I was apprehensive of the opening of a fresh mouth, which might have cut off our retreat. I also feared that the violent explosions would detach some of the rocks off the mountain of Somma, under which we were obliged to pass; besides, the pumice stones, falling upon us like hail, were of such a size as to cause a disagreeable sensation upon the part where they fell.

After having taken breath, as the earth still trembled greatly, I thought it most prudent to leave the mountain, and return to my villa; where I found my family in great alarm at the continual and violent explosions of the volcano, which shook our house to its very foundation, the doors and windows swinging upon their hinges. About two o'clock in the afternoon another lava forced its way out of the same place from whence came the lava last year . . . so that the conflagration was soon as great on this side of the mountain as on the other which I had just left.

The noise and smell of sulphur increasing, we removed from our villa to Naples; and I thought proper, as I passed by Portici, to inform the Court of what I had seen; and humbly offered it as my opinion, that his Sicilian Majesty should leave the neighbourhood of the threatening mountain. However, the Court did not leave Portici till about twelve of the clock . . . I observed, in my way to Naples, which was less than two hours after I had left the mountain, that the lava had actually covered three miles of the very road through which we had retreated. It is astonishing that it should have run so fast; as I have since seen, that the river of lava in the Atrio di Cavallo, was 60 and 70 feet deep, and in some places near two miles broad.

The lava flows must have been unusually hot and fluid to travel at a speed of 2.4 km an hour, although the steepness of the slope over which they travelled would have helped keep them mobile. Nevertheless, most able-bodied persons could, of course, have easily outstripped them. When flows extend 5 km from their source, their edges and surface have usually solidified and they have slowed down to a tenth of that speed. At the same time, however, the eruption was not without explosive vigour. The explosions blew open doors and windows in the royal palace at Portici, and the blasts carried as far as Naples itself, where they blew open the windows in the Palazzo Sessa. To add to the menace, the ground rumbled, hissed and crackled continuously for five hours into the night. The Court had done well to accept Hamilton's advice.

The Neapolitans behaved as they usually did on these occasions and crowded into the streets as if their last hour had come. "His Sicilian Majesty's hasty retreat from Portici added to the alarm; all the churches were opened and filled; and the streets were thronged with processions of saints . . . to quell the fury of the turbulent mountain".

On Tuesday 20 October, the tormentor disappeared behind a pall of ash and smoke that spread as far as the city itself. Fine ash fell on Naples throughout the day, and the Sun appeared "as through a thick London fog, or a smoaked glass". The lava flows on both sides of the mountain were still running very quickly down slope. The falling ash seems to have muffled the noise of the explosions during the day, but the subterranean rumblings resumed at about 9 p.m. and terrified the Neapolitans for four seemingly interminable hours. "It seemed as if the mountain would split in pieces", and a new fissure did open between the two mouths that had previously formed. This prospect of doom proved too much for the prisoners in the public jail. They attacked and wounded their jailer, and tried to escape, but public-spirited troops thwarted their attempt and thrust them back into custody.

The troops repulsed another form of lawlessness with less success. A panic-stricken mob gathered at the cardinal–archbishop's palace and demanded that he bring out the relics of San Gennaro to save the city. Perhaps His Eminence was more afraid of the mob than of Vesuvius? At all events, he refused their raucous pleas. He was unwise. The mob burned down his gates, but even then, the stubborn cardinal–archbishop held on to his precious relics.

On Wednesday 21 October, fears, tempers and Vesuvius all calmed down, and the advancing lavas spared the royal palace at Portici when they changed course a mere 800 m short of it. However, the tranquillity had induced a false sense of security, because the volcano took on a new lease of life the next day. The thundering noise resumed, ash fell so thickly that people had to use umbrellas or pull down the wide brims of their hats to protect their eyes.

The ash and cinders piled up an inch or more thick on the roofs and balconies and they astonished sailors when they fell on ships far out to sea. Hamilton later reported that a farmer living near Portici "lost eight hogs by the ashes falling into their trough with their food: they grew giddy, and died in a few hours". In Naples, the Sun shone as weakly as a pallid moon. Panic and religious fervour again seized the citizens massing in the streets, and this time they had more success:

> The mob . . . obliged the cardinal to bring out the head of Saint Janu-arius [San Gennaro], and go with it in procession to the Ponte Madda-lena [the bridge], at the extremity of Naples towards Vesuvius; and it is well attested here that the eruption ceased the moment the saint came in sight of the mountain.

However, the rationalist in Hamilton could not resist adding a supple-mentary remark in his letter to the Royal Society. "It is true the noise ceased at about that time, after having lasted five hours, as it had done the preceding days".

During the next four days, the volcano began to give off fine ash that was as white as snow. Old timers declared that this was a sure sign that the erup-tion was drawing to a close. The old timers were correct. By then, the snout of the lava flow had extended 10 km from its source. Lava lay over 15 m deep in the valley of the Atrio del Cavallo and it was to stay hot for many months afterwards under its carapace of solidified and blackened clinker. On 19 December, for instance, Hamilton took his old school friend, Lord Stormont (the British ambassador to Vienna) up to see the volcanic sights. The sticks they thrust into the crevices in the flow caught fire at once.

At the end of 1767, Hamilton concluded his survey of the eruption by sending five chests of volcanic samples to the British Museum. He was already thinking of broadening the scope of his stimulating new hobby. Vesuvius, he had discovered, was not the only volcano in Campania.

The volcanoes of the Campi Flegrei and Etna

The Campi Flegrei were full of volcanoes that past eruptions had thrown up like so many giant molehills, just as surely as they had built Monte Nuovo in 1538. In fact, Campania was made up of nothing else but volcanoes as far as the Apennines and the Sorrento Peninsula, and so were all the islands in the Bay of Naples – well, apart from limestone Capri that is. Hot springs, too, occurred all over the region, especially at La Solfatara, and they had been used for medicinal cures ever since antiquity; other springs, such as the Grotto del Cane near Agnano, gave off noxious fumes. South of Monte Nuovo lay the boiling spring known as Nero's Bath; its waters, as Hamilton delicately put it, were "reckoned a great specific in that distemper which is supposed to have made its appearance at Naples before it spread its contagion over the other parts of Europe" [Hamilton could not bring himself to specify that the distemper was venereal disease].

Once Hamilton saw that Campania was so full of volcanic remains, he confessed "that we are apt to judge of the great operations of Nature on too confined a plan". He knew then that he could no longer limit his attentions to Vesuvius. He extended his observations to the rest of Campania and southwards to Etna, the great volcano in Sicily. Hamilton and the young Lord Fortrose, who was out on the Grand Tour, went off to Etna. On 26 June 1769 they were on the summit of Etna with Canon Recupero, the leading expert on the volcano. Although Etna was erupting only fumes, Hamilton gained valuable insights into the way that volcanoes operate. The dozens of old cones and craters on the broad flanks of Etna, the variety of their volcanic forms, and their manifestly different ages, led Hamilton to deduce that the whole of the cone of Vesuvius had been formed by repeated eruptions just like those he had recently witnessed.

Hamilton then took his analysis a stage further and examined the volcanic layers underlying Campania and the ancient soils that were sandwiched between them. Pompeii itself had been built on volcanic rocks that Vesuvius had erupted previously. In AD 79, Pompeii and Herculaneum in turn had been buried by ash and pumice, which by the eighteenth century had already weathered down to a fertile soil almost 60 cm deep. Throughout Campania, outcrops of these older volcanic layers showed that they were interspersed with layers of old soils that had developed on top of an erupted layer during the ensuing calm period; and then the next eruption of ash and pumice had buried these old soils in turn. Thus, the history of a volcanic region could be interpreted by studying the volcanic lavas and fragments, but also by analyzing the ancient soils between them. The number of soils and layers of fragments implied that the volcanoes had been active for a long time, and much longer than the period envisaged by the biblical scholar, James Ussher, who had calculated that the Creation had occurred in 4004 BC.

Campi Phlegraei

The 1770s were the most successful decade of Hamilton's career. He and his wife settled into a glittering social life in Naples and entertained an increasing number of rich and distinguished visitors, drawn to the capital and Vesuvius, at least in part, by Hamilton's own letters to the Royal Society of London. Neither were they disappointed, for ever since 1767 Vesuvius had been giving off fumes and the occasional explosion of cinders. There was even a rare royal accolade from his old schoolboy friend: on 15 January 1772, King George III awarded him the Order of the Bath and he became Sir William Hamilton. Intellectual rewards came too. In London at the end of September 1772, Thomas Cadell published a collection of Hamilton's letters to the Royal Society (with explanatory notes by the author) under the title, *Observations on Mount Vesuvius, Mount Etna, and other volcanos* (sic). A second edition followed in 1773, and a third in 1774.

And yet, these editions were mere stepping stones to the creation of a book of an entirely different scale. Sir William was keen to depict the volcanic landscape of Campania for those who could not see its beauty at first hand and he decided to publish a collection of his letters to the Royal Society in a large book, *Campi Phlegraei: observations on the volcanos* (sic) *of the Two Sicilies*. He commissioned Pietro Fabris, who was perhaps the best of many local artists in the field, to illustrate the volcanic landscape and the rocks in the area. Fabris produced 54 stunning hand-coloured "prints imitating drawings", which have an ethereal quality and delicacy reminiscent of Giambattista Tiepolo, perhaps the greatest Italian painter of the century. Published in 1776, *Campi Phlegraei* was one of the most luxurious volumes produced in the eighteenth century, and it is surely beyond doubt the most beautiful book ever produced on volcanoes. As if to encourage Sir William's glorious initiative, Vesuvius entered into activity from the autumn of 1778, and it reached a spectacular climax during the summer of 1779. Sir William wrote a long letter to the Royal Society about that eruption, and then included it, along with six additional illustrations by Fabris, in a supplement to *Campi Phlegraei*, which he published in 1779.

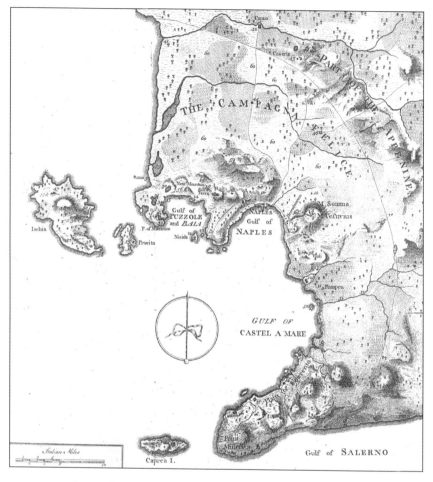

A map of Campania in the eighteenth century (Hamilton 1771).

The eruption of 1779

Sir William Hamilton was not alone in his volcanic studies: Naples had a thriving scientific society at the time, among which Gaetano De Bottis and Giovanni Maria Della Torre were the most prominent members. However, Sir William's work reached a wider audience through the Royal Society of London.

In May 1779, Sir William climbed to the summit of Vesuvius for the 58th time and he had already made a couple of hundred trips to the flanks of the volcano. He passed the night high up on the volcano with Bartolomeo Pumo and a British visitor, a Mr Bowdler of Bath. They saw a typical lava flow leaving its source and running "gently on, like a river that had been frozen, and had masses of ice floating on it". Bartolomeo proposed what at first must have seemed to them like a circus trick – that they should walk across the flowing lava, which "to our astonishment, he instantly put in execution, and with so little difficulty, that we followed him without hesitation, having felt no inconveniency than what proceeded from the violence of the heat on our legs and feet; the crust of the lava was so tough, besides being loaded with cinders . . . that our weight made not the least impression on it, and its motion was so slow, that we were not in any danger of losing our balance and falling on it." However, in his subsequent letter to the Royal Society, Sir William felt duty-bound to emphasize that "this experiment should not be tried except in cases of real necessity . . . should [for instance] anyone have the misfortune to be inclosed between two currents of lava".

Thus emboldened, the trio climbed right up to the source of the flow, which lay about 400 m from the crater:

> The liquid and red-hot matter bubbled up violently, with a hissing and crackling noise . . . like an artificial firework, and by the continual splashing up of the vitrified matter, a kind of arch or dome was formed over the crevice from whence the lava issued. It was cracked in many parts, and appeared red hot within, like an heated oven.

They then climbed up to the rim of the main crater itself, where, in the midst of the shallow crater, a "little mountain" was exploding cinders and clots of red-hot lava. Vesuvius did not make them welcome: "the smoke and smell of sulphur was so intolerable that we were under the necessity of quitting that curious spot with the utmost precipitation".

Throughout July 1779, the usual portents of a more serious eruption became increasingly evident, with noises rumbling underground as well as occasional loud explosions expelling smoke, ash and red-hot cinders. By the end of the month, Vesuvius was providing a beautiful firework display every night. Early in the afternoon of Thursday 5 August 1779, Sir William observed from his villa at Posillipo that Vesuvius had broken into a "most violent agitation". It was expelling puff after puff of smoke like "bales of the whitest cotton", which soon rose at least four times as high as the mountain. Through the midst of these clouds "an immense quantity" of stones, cinders

The eruption of Vesuvius seen from the jetty in Naples on the evening of 8 August 1779, from Campi Phlegraei *(Supplement: 1779).*

The eruption of Vesuvius seen from the jetty in Naples on the morning of 9 August 1779, from Campi Phlegraei *(Supplement: 1779).*

and ash was shooting up at least 600 m. The noise was frightening, even as far away as Posillipo. Another "final eruption" was under way. At 2 p.m., Vesuvius seems to have given off a small pyroclastic flow:

> An extraordinary globe of smoke, of a very great diameter, was distinctly perceived, by many of the inhabitants of Portici, to issue from the crater of Vesuvius, and proceed hastily towards the mountain of Somma, against which it struck and dispersed itself, having left a train of white smoke, marking the course it had taken: this train I perceived plainly from my villa, as it lasted some minutes; but I did not see the globe itself.

The northern slopes of Vesuvius faced other hazards too. Reddish ash fell so thickly upon the towns of Somma and Ottaviano that visibility was restricted to 3 m. Thin filaments like spun glass were mixed with the ash and they excited some curiosity. But a more pressing worry proved to be the nauseous sulphurous fumes and the white salts that covered the trees and burned off their leaves in the surrounding countryside.

At midnight on Saturday 7 August 1779, Vesuvius had its "second fever fit" of the eruption. A summer storm had blown up, but the volcano was not to be outdone. A dazzling fountain of molten lava shot up from the crater and rose "to an incredible height, casting so bright a light, that the smallest objects could be clearly distinguished at any place within six miles or more of Vesuvius". Sir William and all the spectators on the Molo jetty in Naples gaped in admiration. But, to their chagrin, the lava fountaining ceased after only ten minutes; the storm clouds hid Vesuvius from view, and heavy rain began to fall. It was later discovered that the spectacular display had enlarged the crater and obliterated the little cone that had grown up within it.

On Sunday 8 August, Vesuvius was quiet until 5 p.m., when massive fumes gathered over the summit, and explosions of ash and cinders became more and more frequent. At about 9 p.m., a huge blast shook the houses, cracked the walls and broke many windows in and around Portici. Vesuvius was ringing the changes of its volcanic threats: first a lava flow, then the terrible lava fountain, and now a great blast. The eruption was changing from moderate to vigorous.

The great blast was but a prelude to something even worse. Soon, great fountains of dazzling red-hot lava gushed out from the crater along with puffs of pitch-black clouds riddled with silvery-blue forked lightning. The column

soared to such a height that all who saw it could only stand and stare in apprehensive amazement. Connoisseurs of the current aesthetic fashion for the
sublime, like Sir William and his friends, realized that it was "a scene so glorious and sublime as, perhaps, may have never before been viewed by human
eyes, at least in such perfection". Those who also kept their wits about them
calculated that the glowing column was soaring to more than 4000 m, three
times the height of Vesuvius. Most of the drops of lava fell back, still red hot,
and formed a glowing carpet stretching from Vesuvius to Monte Somma,
where they set the brushwood alight. Then, a black cloud flashing with lightning spread out from the crest of the lava fountain and turned towards Naples
itself. Panic. As Sir William recounted:

> All public diversions ceased in an instant, and, the theatres being shut,
> the doors of the churches were thrown open. Numerous processions
> were formed in the streets, and women and children with dishevelled
> heads, filled the air with their cries, insisting loudly upon the relics of
> Saint Januarius [San Gennaro] being immediately opposed to the fury
> of the mountain: in short, the populace of this great city began to display
> its usual extravagant mixture of riot and bigotry, and, if some speedy
> and well timed precautions had not been taken, Naples would perhaps
> have been more in danger of suffering from the irregularities of its lower
> class of inhabitants, than from the angry volcano.

Luckily for the Neapolitan upper classes, the southwesterly wind halted
the progress of the cloud, and the lava-fountaining itself suddenly stopped
after half an hour. Thereafter, "Vesuvius remained sullen and silent. After the
dazzling light of the fiery fountain, all seemed dark and dismal, except the
cone of Vesuvius, which was covered with glowing cinders . . . from under
which, at times, here and there, small streams of liquid lava escaped and rolled
down the steep sides of the volcano". And a smell like a mixture of sulphur
and an iron foundry spread through eastern Naples.

Meanwhile, around Ottaviano, on the northeastern side of the volcano,
the inhabitants had neither the time nor the inclination to admire the view.
The ridge of Monte Somma hides Vesuvius from Ottaviano, so that it was
only at about 9 p.m. on Sunday 8 August that the citizens saw the glowing
volcanic fragments soaring from Vesuvius, high above the ridge. Many rushed
into the churches; others made hasty preparations to leave town. Suddenly,
Vesuvius gave out its great blast and, in a trice, a thick cloud of smoke and

dust enveloped the town. Lightning forked back and forth. "And the sulphureous smell and heat would scarcely allow [the inhabitants] to draw their breath". Another blast heralded an onslaught of cinders and boulders that sometimes reached up to 2.5 m across. Their solid crusts enclosed an interior of molten lava, which splattered out in great showers of red sparks when they crashed to the ground. Straw huts in the vineyards, rooftops and a wood-storage yard in the town soon caught fire. Luckily, there was no wind, otherwise the whole town would have gone up like a tinder box.

The people of Ottaviano were lucky to escape, because the volcanic bombardment was too dangerous for them to leave their homes without risking severe injury. Those who tried to escape ventured out with "pillows, tables, chairs, the tops of wine casks, etc. on their heads, [and] were either knocked down, or soon driven back to their close quarters under arches, and in the cellars of their houses". Many were wounded and two people were killed. This volcanic blitz lasted for 25 minutes before Vesuvius calmed down. The people of Ottaviano did not wait for Vesuvius to resume hostilities. They "deposited the sick and the bedridden, at their own desire, in the churches", and departed while they still had the chance. It is possible that those abandoned in the sanctuaries did, indeed, believe that they, too, would be saved.

At about 9 a.m. on 9 August 1779, the usual rumblings and explosions prefaced the volcano's fourth "fever fit". King Ferdinand and Queen Maria Carolina wisely sent for Sir William to come out to their palace at Posillipo and give them a commentary on the course of events. Clouds of cotton-white smoke billowed forth, sometimes shot through with masses of black ash. The explosions were more violent, and the noise more fearsome, than ever. Huge boulders bombarded the flanks of the volcano. Many citizens in the coastal towns around Vesuvius needed no persuasion to leave for Naples, where "a prodigious concourse of people" paraded the relics of San Gennaro to the Maddalena Bridge. Whether by coincidence or not, the explosive phase waned at about 2 p.m. and lava flowed from the cone for the next three hours. By 7 p.m. the volcano had quietened down again.

At about 5 a.m. on Wednesday 11 August, Vesuvius started its fifth, final and most vigorous "fever fit", which lasted for 12 hours and reached its climax at mid-day with colossal white clouds. During the next two days, Vesuvius gave off nothing but masses of fumes. Sir William could not resist a trip to see what had happened on the far side of the volcano, and on 15 August, he and Count Lamberg, the Austrian ambassador, went to Ottaviano. They found a scene of desolation and misery. An area 4×4 km was covered by a blanket of ash, cinders and boulders, which increased in thickness nearer the town. Nearly all the fruit and leaves had been stripped from every tree and shrub. Small groups of people were staring in dismay at the half-buried ruins of what had been their homes. Some were sweeping the black ash into little piles. "Others were assembled in little groups, inquiring of their friends and neighbours, relating each other's woes, crossing themselves, and lifting their eyes to Heaven when they mentioned their miraculous escapes".

On 18 September, Sir William and his guide Bartolomeo climbed Vesuvius with young Lord George Herbert, the son of his old friend, the Earl of Pembroke. The volcano was still fuming and rumbling profusely, and the crater was 75 m deep. They were lucky that the volcano did not erupt more violently while they were inspecting it. The great lava fountains had splattered back down onto Vesuvius and encased the cone in a shell of solid lava clots that reached 30 m thickness on its northern side and lay over 75 m deep in the Atrio del Cavallo at its foot. Sulphurous fumes issued from every fissure and had tinged the rocks with yellow crystals. Some fragments looked like filaments of glass, and many cinders had a glassy sheen, like obsidian. Here and there, huge boulders thrown from the crater lay scattered around the foot of the cone. This had been a powerful eruptive spasm.

Within the devastated area it seemed as if the vegetation could never spring back to life, but when Sir William returned there on 29 September 1779 the vegetation had recovered, the vines had grown fresh leaves, the apple, peach, pear and apricot trees were blossoming, and some had already formed fruit the size of hazel nuts. Although many foxes, hares and other game had been killed, some survived with the fur singed off their backs – only to succumb to a subsequent royal shoot.

King Ferdinand proved more kindly disposed towards the human victims in the area. As Sir William put it in his letter to the Royal Society in October 1779, "His Sicilian Majesty, whose goodness of heart inclines him on all occasions to show his benevolence and assist the unfortunate, has ordered a considerable sum of money to be distributed among the unhappy sufferers of Ottaviano and its neighbourhood" . . . shades of the Emperor Titus.

After this eruption, Sir William considered that it would be scientifically useful to make a plot of all the changes of the summit area of Vesuvius as they occurred. He encouraged Father Antonio Piaggi, who lived in sight of the volcano near Resina, to record them systematically. Sir William praised him as an "ingenious monk, who is as excellent a draughtsman as he is an accurate and diligent observer". His notes and sketches, covering the period from

From Catherine to Emma

By 1780, Sir William was 50 years of age, fit, healthy, an authority in the youthful new fields of volcanology and antiquarian studies, and a successful British envoy, who had the ear of King Ferdinand and the most favoured place at all the royal hunting parties. Then, on 25 August 1782, his wife, Catherine died at the age of 44. In January and February 1783, severe earthquakes ravaged Calabria. To distract himself from his grief, he went to investigate their effects and composed for the Royal Society a graphic account of the events and their aftermath. Only then did he journey to the UK to arrange his wife's burial and take some leave in London. It was there in 1783 that he first met the woman who was to make him more famous than all the articles on volcanoes that he ever wrote.

Emma (or Amy, Emy or Emily) Hart (or Lyon) was probably born in Cheshire on 26 April 1765. In about 1780, she moved to London, where she worked in what seems to have been an expensive brothel. By all accounts, she was an extremely beautiful young woman, although she was soon to become decidedly chubby. In 1782, Emma became the mistress of Sir William's nephew, Charles Greville. Thus, she met Sir William, who had her portrait painted by both Sir Joshua Reynolds and George Romney. There, for the moment, the relationship rested, and Sir William returned to his post in Naples at the end of 1784. Early in 1785, Greville told Sir William that he was so much in debt that he just had to marry an heiress without delay, and he proposed that his uncle might take Emma off his hands. Sir William succumbed, and Emma and her mother were duly installed in the Palazzo Sessa on 26 April 1786. On a visit to Britain in 1791, the lovers were quietly married in Marylebone parish church on 6 September before they returned to Naples on 1 November 1791. By then, the French Revolution was approaching its climax. On 11 September 1793, as part of the British manoeuvres to counteract French expansion, Admiral Nelson arrived for a brief stay in Naples and met Sir William for the first time. The French revolution was at its height. On 16 October 1793, Marie Antoinette followed her husband to the guillotine. Every court in Europe trembled. Maria Carolina was furious. The hated French had killed her sister. Who better to protect her kingdom than the British?

1779 to 1795, eventually grew into eight volumes, which Sir William presented to the Royal Society of London in 1801.

The eruption of June 1794

In June 1794 came the most vigorous outburst of Vesuvius during the whole of the eighteenth century, and it was to prove to be the longest of all the "final eruptions" that took place between 1631 and 1944. In the preceding years, small eruptions of lava had emerged from near the summit, and had done nothing to dispel the commonly held view that a much greater outburst must be brewing. And yet, Vesuvius had been remarkably quiet for the first six months of 1794, although some local wells had dried up and water levels had dropped in many others. The discharge of the great fountain in Torre del Greco had diminished so much that the water wheels of the cornmill that it supplied turned only very slowly.

Eight days before the eruption, a man and two boys who were working in a vineyard just above Torre del Greco were alarmed when a puff of smoke suddenly exploded from the ground near by (this marked the exact spot from which the main lava flow was later to emerge). Two days before the eruption, the ground rumbled under Resina. As Sir William rightly commented later, if all this information had been communicated at the time "it would have required no great foresight to have been certain that an eruption of the volcano was near at hand, and that its force was directed particularly towards that part of the mountain".

At 11 p.m. on Thursday 12 June, a strong earthquake shook Campania, and especially the coastal towns at the foot of Vesuvius. It lasted for half a minute, starting with an up-and-down motion and ending with an east–west undulation. The flimsier structures all over Campania creaked, cracked and often split open, while the chamber bells rang in the stoutly built royal palace at Caserta. For the next three days, the ground trembled incessantly.

Trinity Sunday, 15 June, was a fine day with no sign of what Sir William called the "electrical fluids in the air" that were then commonly believed to be one of the portents of eruptions. The earthquakes that had shaken the whole area in the vicinity of the volcano for the previous few days had offered a far more reliable warning of the impending outburst. Just after 10 p.m. on 15 June, a huge explosion from Vesuvius shook the ground all around the volcano. A deep and awesome rumbling reverberated through the ground as the magma rose perilously close to the surface. At once, a cloud of very black smoke burst out from near the base of the cone, and an enormous fountain of molten lava followed. More and more fountains followed until at least 15 were spurting from the new fissure. The fissure resembled a livid gaping wound, 1200 m long, slashing down the lower southern flanks of the mountain at a height of between 480 m and 320 m.

Fortunately for the villagers living around the flanks of the volcano, Vesuvius has rarely erupted from low-lying fissures, and most of its activity has been concentrated at or near the summit crater. In 1794, this particular fissure opened a curtain of fire along a line directed straight towards Resina, and a lava flow resembling "the paste of molten glass in a furnace" was soon making its way towards that town.

In his letter to the Royal Society, Sir William put his readers in the picture:

> It was a mixture of the loudest thunder, with incessant reports, like those from a numerous heavy artillery, accompanied by a continued hollow murmur, like that of the ocean during a violent storm; and added to these was another blowing-noise, like that of the going up of a large flight of skyrockets, and which brought to my mind also that noise which is produced by the action of the enormous bellows on the furnace of the Carron iron foundry in Scotland, and which it perfectly resembled. The frequent falling of the huge stones and scoriae [cinders], which were thrown up to an incredible height from some of the new mouths . . . contributed undoubtedly to the concussion of the earth and air, which kept all the houses at Naples for hours in constant tremor, every door and window shaking and rattling incessantly, and the bells ringing.
>
> This was an awful moment! The sky, from a bright full moon and starlight, began to be obscured; the moon had presently the appearance of being in an eclipse, and soon after was lost in obscurity. The murmur of the prayers and lamentations of a numerous populace forming various processions, and parading in the streets, added likewise to the horror . . . I recommended to the company . . . who began to be much alarmed, rather to go and view the mountain at some greater distance, and in the open air, than to remain in the house, which is . . . in the part of Naples that is nearest and most exposed to Vesuvius. We accordingly went to Posilipo [sic] and viewed the conflagration, now become still more considerable, from the sea side under that mountain.

That night, the Neapolitans were so afraid of a major earthquake that few dared sleep in their beds. (This fear of earthquakes seems to be a common feature when cities are threatened by volcanic eruptions. It is rarely justified and the eruptions themselves can sometimes prove far more devastating.) In fact, the only earthquakes felt distinctly in Naples itself occurred on 12 and 15 June, although near Vesuvius the ground shook incessantly, with some more powerful bursts often associated with major emissions of lava. As Sir William reported, the Neapolitans ran true to form: "The common people were either employed in devout processions in the streets, or were sleeping

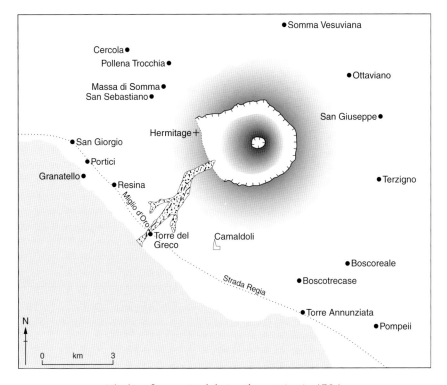

The lava flows emitted during the eruption in 1794.

on the quays and open places; the nobility and gentry, having caused their horses to be taken from their carriages, slept in them in the squares and open places, or on the high roads just out of town".

At about 2 a.m. on 16 June, Sir William observed that lava was then flowing freely from the fissure. He therefore concluded that Naples was no longer in danger from an earthquake, which had been his greatest fear, and that the Hamilton household could return to the Palazzo Sessa. However, although "there was not the smallest appearance of fire or smoke from the crater on the summit of Vesuvius", the black smoke and ash issuing from the fissure formed a dense body of clouds that rose above the mountain. Thus it was that, in the third "final eruption" after those of 1767 and 1779, Sir William once again saw Vesuvius develop the soaring lightning-riddled cloud with the pine-tree shape, which Pliny the Younger had made famous in AD 79. But the eruption was far from over. The destruction was yet to come.

At about 4 a.m. on 16 June, Vesuvius began to explode vast clouds of fumes that hid the summit for most of the next few days. At times, loud

215

The eruption of 1794 seen from Naples, after a painting by Alessandro D'Anna. The lava flow that invaded Torre del Greco can be detected on the profile of Vesuvius that forms the horizon on the right.

explosions were quickly followed by a red tinge to the clouds, which showed that molten fragments were being ejected. At dawn on 16 June, a lava flow issued from the northeastern side of the volcano above Ottaviano and set a wood alight before it halted at the garden plots on the fringe of the town.

However, the clouds of fumes did not hide the main disaster from the eyes of the spectators who had gathered at daybreak at every vantage point in Naples. They watched open mouthed as the lavas continued to gush from the fissure and flow towards Resina and perhaps even towards Naples itself. Then, however, at Tironi, the snout of the advancing flow entered a little ravine that served to divert the lavas seawards, away from Resina and towards Torre del Greco.

The relief in Naples and Resina matched the horror among the 18000 citizens in Torre del Greco. Just at that moment, too, a new flow burst out in a vineyard only 1.5 km from the town, destroyed the vines, joined the main stream and gave it even greater vigour. Some thought to try to divert the flow from its new course, but the fluid lavas were moving much too fast for any workable plans to be implemented in the confusion of the night. Within about an hour, the combined flows streamed in a broad front straight through the very centre of Torre del Greco, igniting, swamping, and burying

The campanile of Santa Croce in Torre del Greco after the eruption in 1794. The campanile and the church now have an unusual architectural relationship. In 1794, the lavas destroyed and buried most of the church of Santa Croce, but did not cover the upper parts of the higher red-brick campanile. A new (white) church was built on top of the thick solidified lava flow and it thus now rises higher than the old campanile.

– almost, it seemed, digesting – its finest buildings, as well as four fifths of the town, as they surged onwards to the sea. The lava did not even spare large buildings: for instance, it buried most of the church of Santa Croce, except for the upper parts of its campanile.

Nevertheless, some citizens managed to counteract the indiscriminate assault of the flow. As the lavas surrounded one of the convents, a brave friar managed to rescue five or six old nuns who had a sweet tooth. The friar told Sir William that:

> Their stupidity was such as not to have been the least alarmed, or sensible of their danger: he found one of upwards of 90 years of age actually warming herself at a point of red-hot lava, which touched the window of her cell, and which she said was very comfortable; and though now apprized of their danger, they were still very unwilling to leave the convent, in which they had been shut up almost from their infancy, their ideas being as limited as the space they inhabited. Having desired them to pack up whatever they had that was most valuable, they all loaded

217

themselves with biscuits and sweetmeats, and it was but by accident that the friar discovered that they had left a sum of money behind them, which he recovered for them.

Some justice was also done, as an anonymous local lawyer remarked:

> There was no dearth of desperados, who disdained the fire over which they had to walk in order to gain entry into, and rob, the palaces buried beneath the ash. However, they died horribly in the process because the walls had already been burnt; and as soon as their feet touched them, they were entombed under the falling debris.

The lavas must have been unusually hot and fluid, because they often flowed rapidly where they were channelled along the streets. The wiser citizens of Torre del Greco quickly clambered up to higher ground bordering the central valley where the town had been built. From their vantage points they then watched "the martyrdom of their town as lavas swallowed up the centre". Other citizens decided to leave for more distant sites: those to the east of the flow made for Castellammare di Stabia, while those to the west of the flow rushed along the coast towards Naples. There was no time to move many of the old and infirm from Torre del Greco itself; and 15 of them were left lying in their own homes until the molten lava enshrouded them. Many able-bodied people were also trapped in their homes, but they took refuge in upper rooms, or on the rooftops, desperately hoping that the flow would not rise any higher. They had to wait anxiously until the following day before they dared scramble out across the surface of the solidifying lava.

The hot flow entered the sea at Torre del Greco with an enormous hiss of steam. A huge fountain of sea water sprayed high into the air like a geyser. Boiled fish floated up to the surface of the water. Then, by coincidence, once the flow had reached the sea, the supply of lava waned almost at once. Its spitting snout was then 8 km from its main source and had formed a lava promontory stretching about 200 m out into the sea.

On the morning of 17 June, Sir William rather unwisely took his boat to investigate the scene at close quarters. Filaments of molten lava were still bubbling from the solidifying snout and heating the sea 100 m all around. Sir William scalded his hand when he dipped it into the water in a rare injudicious experiment. That was not all. "By this time, my boatman observed that the pitch from the bottom of the boat was melting fast, and floating on the

TORRESI
UN SECOLO CI CONTEMPLA
DA QUESTO STORICO
CAMPANILE
RESISTETTE IMMOTO
ALLA IMMANE ERUZIONE
DEL 15 GIUGNO 1794
FU SPRONE
ALLA COSTRUZIONE DELLA NUOVA
CITTÀ
COMMEMORANDO OGGI
TALE SOLENNE AVVENIMENTO
INCHINIAMOCI ALLA VIRTÙ
DEGLI AVI
PROCURIAMO DI RENDERE
IL NOSTRO PAESE
SEMPRE PIÙ PROSPERO E BELLO

IL MUNICIPIO
TORRE DEL GRECO 15 GIUGNO 1894

The inscription to commemorate the centenary of the eruption:
"People of Torre del Greco, a century looks down upon us from this historic campanile, which stood firm against the immense eruption on 15 June 1794 and provided the incentive for the construction of the new city. Today, as we commemorate that solemn occasion, let us bow to the virtue of our forefathers and let us work to make our land ever more prosperous and beautiful. The Municipality, Torre del Greco 15 June 1894."

surface of the sea, and that the boat began to leak. We therefore retired hastily from this spot, and landed at some distance from the hot lava".

Sir William climbed to the upper part of the town for an overview of the devastation. The lava flow was about 1600 m wide and up to 12 m thick in the valley that passed through the centre of Torre del Greco. It had buried many buildings to their upper storey, and some of their timbers were still on fire. From time to time, explosions sent ash and cinders flying into the air when small amounts of gunpowder (commonly kept in houses) were set alight. In fact, it soon turned out that a disastrous explosion had been narrowly avoided. Many fireworks had been stocked in a warehouse for a forthcoming fiesta. It was just as well that the lava did not reach them, or many citizens might have suffered a sudden and unexpected death. Sir William confessed that "I should not have been so much at my ease had I known of this gunpowder", for even after the flows stopped, the temperature in the town was "very near a hundred degrees" (about 39°C). In a perilous gesture, the firework maker saved his investment, and the populace, by carrying his dangerous goods to safety over the still-hot lava. Sir William also witnessed a facet of "Happy Campania" that was as depressing as the effects of the lava: a few of the citizens, whose homes had survived, had returned to recover their belongings; to "their cruel disappointment", they found that everything of value had already been looted from their houses.

In the end, it was a man with a pig who finally persuaded Sir William to retreat from "this melancholy spot". One of these robbers came running towards Sir William carrying a pig on his shoulders. The pig's rightful owner was pursuing him, armed with a shot gun. To avoid being shot, the robber began to dodge around Sir William. The envoy kept his presence of mind, adopted his most aristocratically British tone, and bade the robber to drop the pig and depart. The robber, of course, obeyed. The owner of the pig was so grateful that he warned Sir William that there were thieves in every house. He therefore decided that it would be safer for both himself, and the dispossessed citizens of Torre del Greco, if he were to go to Naples and find some protection for them. Once Sir William had recounted his adventures, the authorities immediately sent troops to the town, but they had to travel by sea because lava had blocked the road.

While Sir William was in Torre del Greco, fine ash fell on Naples throughout 17 June, hiding the volcano and bringing darkness to the city. There was the customary Neapolitan reaction: nothing was heard in the streets except the explosions from the volcano and the prayers from the parading penitents.

At 4 a.m. on Wednesday 18 June a strong earthquake shook the area around the volcano and induced many people to run out into the streets. For a while after daybreak, the wind cleared the clouds from Vesuvius and revealed that much of the crater facing Naples had fallen in. Then yet another huge cloud of smoke and ash "looking just like a cauliflower" towered high over the summit, expanding all the while. The ash rained down and coated the whole area around the volcano in melancholy light grey. Dense ash "as fine as Spanish snuff" continued to shower down on the city. Sir William had been depressed by what he had seen at Torre del Greco and he viewed the prospects of Naples with pessimism. He feared that Naples might be about to suffer the fate of Pompeii, because the ash seemed to be very much like that he had seen in the excavations there. But, impressive though it was, the eruption in 1794 was several degrees of magnitude lower than the outburst in AD 79. In any case, his fears proved to be groundless when the wind changed direction and blew the ash inland.

Meanwhile, from 16 to 21 June, on the northern slopes of the mountain, black ash and fiery cinders rained down, turned day into night, and forced the villagers to live by candlelight. Those with any forethought shovelled the ash away from their rooftops, although the roof of one house in Somma village caved in and killed the three unfortunates inside. It was even claimed, with more than a little poetic licence, that the ash fell in such torrents that people had to keep moving so that they would not be rooted to the spot. The villagers did not put that view to the test. They ran off, desperately trying to keep their torches alight so that they could see their way along the road.

The main outburst was over by 22 June, although sporadic activity continued until 7 July. However, from the last week in June, mudflows proved to be the greatest source of danger to the towns around the volcano. These mudflows formed whenever heavy rains drenched and mobilized the thick blanket of ash covering all the flanks of Vesuvius. The result was that water, mud, ash, boulders, masonry and uprooted trees flooded down the slopes, destroyed the rich crops and swept thousands of sheep and cattle, and even a team of eight oxen, to their deaths. Several mudflows swept seawards between Torre del Greco and Torre Annunziata on 7 July. On the northern slopes of Monte Somma it was like 1631 all over again. Especially on 12 July, major mudflows ran down to Somma and Ottaviano and caused more serious and prolonged devastation than the blanket of ash, because the mud dried like brick, which could then be cracked open only with pick axes. On the other hand, given the appropriate muscle power, the ash was easy to remove from homes and streets, and it usually became fertile within a year if it had been thin enough to be left on the fields.

Sir William delayed his next expedition to Vesuvius until 30 June, when he made his 68th ascent to the summit with his guide, Bartolomeo Pumo. They saw "nothing but ruin and desolation". The weight of ash had broken branches from the trees and either buried the vines or burned off all their leaves. At the base of the mountain, the pale-grey ash was 30 cm thick, but nearer the cone it was at least 3 m deep, and it completely masked the rugged surfaces of all the most recent lava flows, so that, for once, they were easy to cross. The lavas of the great destructive flow of 15 June were still hot under their blanket of ash. Where the lava fountains had gushed out along the fissure, they had piled up cones of clots of lava nearly 50 m high, but their gaping craters were 180 m or more deep. And here and there along the fissure, sulphurous fumes were still swirling from a multitude of small holes. The main cone of Vesuvius was now about 120 m lower than before the eruption. The crater at its crest was now 156 m deep and 712 m in diameter. However, calm soon returned, for when Scipione Breislak climbed to the summit, on 12 July, he noted "the perfect tranquillity of the volcano".

On the other hand, the lower fissure was not yet entirely benign and several craters blasted out fine ash on 15 July, 22 July and 15 August, after water had filtered down to the magma only to be violently converted to steam. Meanwhile, invisible noxious gases, such as carbon dioxide, which are denser than air, still accumulated in cellars or in low-lying areas in the countryside. They killed hundreds of hares, partridges, pheasants and cats, for instance, in the royal hunting grounds. However, one ingenious farmer near Resina, whose vines had been poisoned in 1767, was better prepared in 1794. He dug a deep ditch around his vineyard: the poisonous gas drained into the ditch and his own vines remained unharmed while those of his neighbours perished.

Some commentators declared that the eruption had cost no less than 500 lives. More sober estimates counted 15 bed-ridden victims in Torre del Greco (who had been abandoned by their relatives or neighbours), three people killed by collapsing roofs in Somma, and two in Ottaviano, while carbon dioxide gases asphyxiated three people in Resina. Thus, the eruption claimed 23 known human victims. To these might possibly be added a few deaths reputedly caused by the lightning that played around the volcano almost throughout the eruption. Ash had fallen as far as Apulia and Calabria, damaging crops and trees, so that the countryside looked as if it was in the depths of winter. Several hundred farm animals met their deaths, either directly through the action of the ash or fumes, or indirectly because the ash covered grasslands and fodder so that they starved. In the arable areas, the crops were ruined for two years.

The destruction of much of Torre del Greco brought out the best in King Ferdinand. In fact, the king was as genuinely kind to his subjects as he was merciless to the beasts of the field. As one anonymous writer observed:

> Our king opened up his treasury to comfort the distressed citizens of Torre del Greco; and, to that end, he appointed the distinguished knight, Don Vespasiano Macedonio, and the most prudent and successful lawyer, Don Angelo Paduano. As soon as they arrived there, they had the ovens opened, and had all other assistance implemented to prepare food for all these people, who had been rich the previous evening and were now left humble, homeless and completely ruined.

The king then offered the citizens a safer spot where they could rebuild their town. Torre del Greco lies in one of the most vulnerable positions around Vesuvius: it had already been badly hit in 1631, and lava had destroyed its eastern sectors in 1737. Yet, none of the citizens opted to accept the alternative site that Ferdinand proposed. Like most victims of volcanic or seismic disasters, the people of Torre del Greco were too attached to their native soil to move elsewhere. Thus, they set about rebuilding their homes and their lives before the invading lavas had even cooled.

Enter Nelson

Meanwhile, the French and the British were at war again. On 1 August 1798 Admiral Horatio Nelson won the Battle of the Nile and, on 22 September 1798, he returned to Naples on his ship *Vanguard* as a fêted hero. Naturally, he stayed with the proud British envoy at the Palazzo Sessa. That autumn, the French army invaded Italy. The Neapolitan court panicked, but Sir William and Nelson organized the flight of the royal family to their Sicilian capital, Palermo, with their bags of gold and silver plate. The French entered Naples, and sympathizers set up a republic on 25 January 1799. It did not last long and, when the royal family returned to Naples, Nelson was in the forefront of the merciless revenge wreaked upon those who had supported the French. He was also coming to the forefront of Lady Hamilton's affections.

Thus began perhaps the most famous *ménage à trois* of the century. Nelson became a laughing stock in the fleet. Sir William became a figure of fun in high society, his admirable career tumbling to a ludicrous end. He left Naples in 1800, never to return. In November 1800, the trio settled in London. On 6 April 1803, Sir William Hamilton died in Emma's arms, with Nelson holding his hand. Nelson died a famous death at Trafalgar on 21 October 1805. Emma herself died in lonely poverty in Calais in January 1815.

Hamilton as a volcanologist

Sir William Hamilton probably knew nothing about volcanoes when he arrived in Naples in 1764, but he had become one of the founders of modern volcanology when he left the city in 1800. The greatest value of his work lay in his acute and careful observations and in his graphic clear descriptions and analyses of eruptions in his letters to the Royal Society, while his glorious book *Campi Phlegraei* did much to broaden interest in volcanoes.

As a background to his own observations, Hamilton seems to have read the works of the classical authors and those of eighteenth-century authorities such as Serao and Della Torre in Naples, Buffon and Desmarest in France, and Hutton and Playfair in Scotland. He maintained a humble modesty about the value of his own studies and he felt unable to develop a system or methodology about volcanoes. "I shall content myself with collecting facts", he wrote, "and let who will form them into a system afterwards". He formulated no grand schemes, such as the crazy theory developed by Leopold von Buch, which perverted volcanology for decades – that volcanoes were formed by the prior elevation of the ground, like so many boils on the skin.

At that time, many students of the Earth still believed that volcanic eruptions could occur from any mountain. However, it soon became obvious to Hamilton, and to all those who actually looked at eruptions, that volcanoes had been formed when volcanic fragments were thrown from, and then accumulated around, a volcanic vent. "The formation of such conical mountains with their craters", he wrote, "are [sic] easily accounted for by the fall of the stones, cinders, and ashes emitted at the time of an eruption". He thus recognized that "mountains are produced by volcanos [sic], not volcanos by mountains". The eruptions themselves had built the volcanic mountains. This simple and easily verifiable fact was not to be universally acknowledged for several decades.

These eruptions produced a characteristic landform wherever they occurred and they served to distinguish volcanoes from other mountains. Therefore, volcanoes could be identified whenever this particular shape was observed in the landscape, even where volcanic eruptions had never been seen or suspected. He agreed with other students of volcanoes that the remains of ancient volcanoes could still be identified, in such regions as the Rhineland in Germany and Auvergne in France. "Whenever I shall meet with a mountain, in any part of the world, whose form is regularly conical, with a hollow crater on its top, and one side broken, I shall be apt to decide such a mountain's having been formed by an eruption".

Hamilton was one of the first to study volcanoes by watching closely how they erupted. Thus, he saw that molten lavas solidified and formed layers of rocks that could be seen not only in present-day volcanic areas but also in areas where volcanic activity had ceased. He saw, too, that the layers of pumice and ash clothing much of Campania had also been erupted as fragments from the volcanic vents that marked the whole region.

His studies of Vesuvius, the Campi Flegrei and Etna quickly convinced him that the Earth must be much older than the timespan determined by James Ussher. As Hutton had so ably demonstrated in Edinburgh, Hamilton also realized that nature was complex and had always acted over a very long period of time. "Nature", he wrote, "acts slowly, it is difficult to catch her in the act". The prolonged and repeated eruptions that formed volcanoes were a strong indication of the great age of the Earth. Excavations in Campania showed that the older volcanic rocks lay under the present landscape. Thick layers of old soils commonly separated volcanic layers, which demonstrated that past eruptions had often been separated by long quiet periods when the surface of the volcanic rocks had been weathered down into fertile soils. Such soils were often thicker than those clothing the present landscape. The number of volcanic layers and the many layers of ancient soils between them implied that the volcanoes had been active for a long time, and thus that the Earth was much older than the biblical scholars had decreed.

The landscape contained many different volcanoes of different ages. Breaking new ground in methodology, Hamilton thus realized that he could distinguish the results of recent eruptions from those of the more distant past. They could be identified by their form, depth of weathering, vegetation cover and the presence of hot springs. More recently formed volcanoes had sharp outlines, clear craters, little vegetation and thin soils. Volcanoes formed in the distant past had blunt outlines, poorly defined craters, and deeply weathered soils that often supported thick and flourishing vegetation. As the volcanic fragments weathered more deeply and were gradually colonized by vegetation, so the volcanic forms themselves lost their original sharp outlines and "mouldered down" until their craters appeared like "a dimple or hollow on their rounded tops".

Moreover, Hamilton soon became convinced that volcanic eruptions were basically constructive and not destructive, in spite of all the calamities associated with them. "Volcanoes should be considered in a creative rather than a destructive light". They added to the land, and rarely accomplished more than superficial destruction. Indeed, the very fact that volcanoes rose above their surroundings demonstrated that constructive forces predominated. Such views provided some of the first steps in developing a method of observing and analyzing a volcanic landscape.

Hamilton also acknowledged one of the fundamental truths of the Earth sciences, the principle of uniformity: that "nature, though varied, is certainly in general uniform in her operations". The processes of nature that were seen today had happened in the same way in the past. The present thus provides the key to what happened in the past.

Further reading

Acton 1956; Breisak & Winspeare 1794; Carta et al. 1981; Constantine 2001; Fothergill 1969; Hamilton 1767–1795; Serao 1743.

Chapter 10

Vesuvius as a tourist attraction:
the Grand Tour

Frequent impressive eruptions made a visit to Vesuvius an enthralling and almost compulsory climax for the young aristocrats on the Grand Tour and did much to establish the fame of the volcano.

Little by little during the late seventeenth and early eighteenth centuries, the notion took hold in northern Europe that the young gentlemen who were destined to become the leaders of society should complete – or in some cases, start – their cultural education by going on a Grand Tour of the classical and Renaissance sites of southern Europe. The excursion commonly lasted up to two years and, gradually, more and more young men returned home with a veneer of knowledge and culture, as well as an impressive collection of souvenirs. Soon, the Grand Tour graduated from good idea to fashionable obligation. The great sites of Italy were a compulsory destination. The cultural treasure houses of Venice, Florence and Rome were obvious early goals, but by the middle of the eighteenth century Naples came to form the glamorous climax to many a Grand Tour because it also possessed two fabulous supplementary attractions: Vesuvius in eruption, and the excavations of Pompeii and Herculaneum. Soon, any self-respecting aristocrat and person with any pretensions to culture just had to go there. The tourists were rich and usually very well connected. One stayed with one's envoy or ambassador; and one had to be seen in the salons of Naples amid a flurry of balls, concerts and visits to artists and dealers in antique statuary. Naturally, one had to make the spectacular excursion to the redoubtable Vesuvius.

The visitors who flocked to Campania included: Addison, Andersen, Berkeley, Berlioz, Casanova, Chateaubriand, the future King Christian VIII of Denmark, Davy, Denon, Dickens, Dumas, Evelyn, Fragonard, Gibbon, Frederick-William III of Prussia, Goethe, the Emperor Joseph II, Humboldt,

Sketch of Vesuvius in 1819, based on an unfinished watercolour and pencil drawing by J. M. W. Turner, and seen from the southwest, perhaps in the vicinity of Posillipo.

Lamartine, Metternich, Milton, Montesquieu, Mozart, the future Tsar Paul I of Russia, Ruskin, Shelley, Madame de Staël, Swinburne, Turner, Madame Vigée Le Brun and Winckelmann. The complete list is so long and the general calibre of the visitors so high that it is tempting to ask what excuse was proffered by those who stayed at home.

Picturesque, sublime and classical

On the Grand Tour, the gentlemen, and a few adventurous ladies, felt obliged to experience the three most important cultural themes of the Age of Enlightenment: the picturesque, the sublime and the classical. In the picturesque, nature showed all its infinite variety; in the sublime, the eruptions showed nature in all its awesome and overwhelming power; and in the classical they believed they saw art at its apogee. There was no better place on Earth than Campania to observe all these themes in vivid combination; and even the greatest numbskulls among the visitors could appreciate the extraordinary variety that the region presented; and here was nature, society and archaeology made life and art. Here too was the most famous volcano in Europe.

Naples still shared the distinction of being considered with London and Paris as one of the three most populous cities in Europe. Naples made Rome

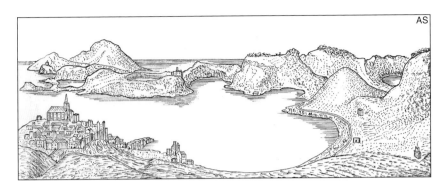

The Campi Flegrei in the eighteenth century. Pozzuoli stands on the headland in the left fore-ground; Miseno lies on the headland in the middle distance on the left, and Ischia forms the horizon on the left. The castle of Baia stands on a promontory in the centre; Monte Nuovo rises between the Bay of Pozzuoli and the hollow containing Lake Averno on the right.

look provincial – a great city with a population that mixed charm and crime, and which seemed to draw some of its unparalleled vitality from the outbursts of volcanic energy that Vesuvius had been regularly displaying on the eastern horizon since 1631. Here was the most vibrant and bustling city in Europe, which had a *joie de vivre* that sent visitors such as Goethe from the staid and sober north reeling with amazement.

Within an easy ride from Naples lay the two most celebrated resurrected cities of the ancient world, which were yielding treasures of classical beauty beyond compare and almost beyond price. Relics of ancient Pompeii became as fashionable as Canalettos from Venice. It seemed that every man of taste must then build a mansion in the classical style, which was not unlike the villas springing up on the lower slopes of the volcano. In Britain, for instance, a gallery decorated in the classical style of Robert Adam could display their souvenirs: an oil painting of Vesuvius in eruption by Volaire or Wright of Derby, a watercolour by Lusieri, a gouache by Fabris, or a view of the ruins by Hackert – not to mention antique vases and statues that venal and thoughtless vandals had pilfered with their pick axes from the most famous ancient cities on Earth.

Naples was just close enough to display the eruptions in such clear and fearsome detail that observers could indulge themselves in exquisite frissons of terror. Those who took the trouble to climb the mountain sometimes earned the added thrill of being roasted by the glowing lava flows, almost stifled by the fumes, and even of being showered occasionally in ash and pumice, or bombed by the rocks and clots of molten lava hurled from the crater. They had all the excitement of the battlefield, all the glamour of the sublime, and all the horror of a glimpse of hell; and then, back they could go to a salon for a sonata on the harpsichord.

In the nineteenth century, when the fashion changed to romanticism and travel became easier, Campania still held some trump cards: strikingly evocative pictures of nocturnal eruptions, romantic stories by the dozen, and also increasingly scientific archaeological investigations, revealed the daily life of the citizens of Pompeii and Herculaneum, and even more works of art. As the fame of the region spread, writers and journalists provided accounts for those in the middle classes who would never be able to afford to visit Naples and Vesuvius themselves. In 1845, Charles Dickens offered a glimpse of the Neapolitan "pickpockets, buffo singers and beggars, rags, puppets, flowers, brightness, dirt, and universal degradation; airing its Harlequin suit in the sunshine . . . singing, starving, dancing, gaming, on the seashore; and leaving all labour to the burning mountain, which is ever at its work . . . the mountain is the genius of the scene . . . the doom and destiny of all this beautiful country, biding its terrible time." Later came the tours organized by travel agents who now fill the sites with obedient camera-laden groups. Today, computer-enhanced television programmes can bring some inkling of the eruption of AD 79 and the great disaster of Pompeii to spectators who never need to leave their sitting rooms.

Until the beginning of the twentieth century, the visitors had to rely almost exclusively on their pens, sketching pads, and their imagination, to describe what they saw. Nowadays, it is hard to appreciate how difficult it was to find words to describe the city, let alone a volcanic eruption, without the benefits of photography; it was a task that would have daunted Shakespeare. The accounts show what persistent moderate activity really looked like at close quarters. The visitors usually witnessed moderate eruptions, where lava flows emerged, hissing, from small vents and fissures on the cone, where lava fountains might spurt high into the air, and where noisy explosions showered down clouds of dust, ash and pumice all over the volcano. To many, the volcanic features were scarcely more than a huge firework display or a fairground spectacle, and their accounts are little more than glorified postcards. The valuable exceptions were the talented observers, whose literary skills and striking comparisons graphically brought the variety of the eruptions and the Vesuvian landscape to life and evoked not only what they saw and heard but also what they felt and imagined. At times, their prose seems purple by our rather impoverished contemporary standards, but they

Major eruptions of Vesuvius, 1631–1794

1631 16–19 Dec.	Violent eruption with pyroclastic flows; villages S and W of Vesuvius badly damaged; 4000+ deaths (SP) (VEI 4)
1637 July – 14 Jan. 1650	Strombolian eruptions, possibly until 1698 (S)
1660 3–29 July	Strong Strombolian eruption: ash and lava flows (I) (VEI 3)
1680 26–28 March	Powerful eruption: ash and lava fountains (F) (VEI 3)
1682 12–22 Aug.	Powerful eruption: widespread ashfalls; 3 deaths (F) (VEI 3)
1685 26 Sept. – 4 Oct.	Explosive eruption (I)
1689 17 Nov. – 15 Dec.	Explosive eruption; main cone 66m higher (I)
1694 5 April – 30 May	Powerful eruption; lava flows spill over the rim of the crater and reach coastal villages to SE and W (F) (VEI 3)
1697 15 Sept.– 9 Jan. 1698	Lava fills crater of 1631 (I)
1698 19 May – 1 June	Powerful eruption; ashfalls, mudflows and lava damage crops (F) (VEI 3)
1699 22 April – 27 July 1707	Persistent Strombolian activity (S)
1707 28 July – 22 Aug.	Powerful explosive eruption; ashfalls, lava flows and mud-flows cause widespread damage (F) (VEI 3)
1712 5 Feb. – 16 March 1730	Persistent Strombolian activity (S)
1712 5 Feb. – 8 Nov.	Sporadic eruptions; lavas towards Torre del Greco (I)
1714 15–30 June	Eruption of lava flows, fountains and ash (I)
1717 6–18 June	Eruption of lava from flank of cone (I) (VEI 2–3)
1720 7–27 May	Explosive eruption of ash over Ottaviano (I)
1721 1 May – 7 June	Lava flow towards Torre del Greco (I)
1723 20 April – 8 June	Powerful eruption of lava flows towards N, followed by eruptions of ash, causing much damage on E of volcano (F) (VEI 3)
1724 4–29 Sept.	Eruption of both ash and lava flows (I)
1730 27 Feb. – 1 April	Powerful eruption of ash and lava (F?) (VEI 3)
1737 20–30 May	Powerful eruption of ash and lava; flows towards Resina and then to Torre del Greco, which was badly damaged (F) (VEI 3+)
1744 1 July – 22 Dec. 1760	Many episodes of persistent Strombolian activity (S)
1760 23 Dec. – 5 Jan. 1761	Powerful eruption from fissure near Boscotrecase: lava flows and six small cones (F) (VEI 3)
1766 23 March – 10 Dec.	Strong eruption, with lava flows and explosions (I)
1767 19–27 Oct.	Lava flows spilled towards Portici and San Giorgio, followed by powerful explosions (F) (VEI 3)
1779 29 July – 13 Aug.	Lava flows, huge lava fountains 4 km high; ash fell on Ottaviano, Somma, Torre del Greco, and Portici; wide-spread damage, three victims recorded (F) (VEI 3)
1783 18 Aug. – 14 June 1794	Episodes of persistent Strombolian activity (S)
1794 15 June – 7 July	Fissures low on SW and NE flanks; 80% of Torre del Greco destroyed by lava flow; ash and mudflows damage Ottaviano and Somma; at least 18 deaths (F) (VEI 3–4)

VEI = volcanic explosivity index (devised to compare the power of different eruptions) S = persistent moderate effusive–explosive Strombolian activity I = intermediate eruption F = final eruption SP = sub-Plinian eruption
N.B. The types of eruption are not as clearly distinct as this table implies.

often conjure up vivid impressions of these extraordinary scenes. Some visi-tors, such as George Berkeley, Sir Humphrey Davy and Vivant Denon, even tried to analyze in a scientific way what they saw at a time when volcanology was still in its infancy.

The view of Vesuvius from Naples

Everyone knew about the spectacular view. It had been one of the most famous sights in the world for centuries. To John Evelyn in 1645, for instance, the view from the Saint Elmo Castle above Naples formed "one of the most divertisant and considerable vistas in the world". Even now, stand at the end of the Via Partenope in Naples, look eastwards across the bay, and the famous phrase springs inevitably to mind: *Vedi Napoli e poi muori* – "See Naples and die". The dictum would be a cliché if it were not inescapable: even Goethe himself could not resist it in 1787.

At night, the eruptions were grandiose and enthralling. On 18 December 1818, Shelley described how, after sunset, "the effect of the fire became more beautiful. We were . . . surrounded by streams and cataracts of the red and radiant fire; and in the midst, from the column of bituminous smoke [that was being] shot up into the air, fell the vast masses of rock, white with the light of their intense heat, leaving behind them, through the dark vapour, trains of splendour".

Goethe, the Duchess and Vesuvius

On 2 June 1787, Johann Wolfgang von Goethe climbed up to the top floor of the royal palace in Naples to pay his respects to the young Duchess of Giovene. It was twilight and the duchess immediately opened the window shutters.

The sight was such as is seen only once in a lifetime. The window . . . was . . . directly facing Vesuvius. The sun had set some time before, and the glow of the lava, which lit up its accompanying cloud of smoke, was clearly visible. The mountain roared, and at each eruption, the enormous pillar of smoke above it was rent asunder as if by lightning, and in the glare, the separate clouds of vapour stood out in sculptured relief. From the summit to the sea ran a streak of molten lava and glowing vapour, but everywhere else, sea, earth, rock and vegetation lay peaceful in the enchanting stillness of a fine evening . . . Though it was impossible to discern every detail, the whole could be taken in at a glance . . . We had before our eyes a text to comment on for which millennia would be too short. As the night advanced, every detail of the landscape stood out ever more clearly; the moon shone like a second sun; with the aid of a moderately strong lens, I even thought I could see the fragments of glowing rock as they were ejected from the abyss of the cone. My hostess . . . had ordered the candles to be placed on the side of the room away from the window. Sitting in the foreground of this incredible picture, with the moonlight falling on her face, she looked more beautiful than ever . . .

The trip to the foot of Vesuvius

Sir William Hamilton escorted the élite to the volcano. In March 1769, for instance, he accompanied the Austrian Emperor Joseph II, the most enlightened monarch in Europe. Unfortunately, the imperial enlightenment did not extend to guides, for Joseph II soundly caned their guide, Bartolomeo Pumo, for some misdemeanour en route. The Russian Grand Duke Paul (who would become Tsar Paul I) caused Sir William a different problem. In February 1782, the heir to Catherine the Great came to Naples with his second wife. Hamilton was enrolled to take the visitors to Vesuvius. Paul spent most of the journey fondling his wife, explaining to Sir William that he loved her very much – after all, they had only been married since 1776. Sir William later recounted that:

> Their imperial highnesses were quite knocked up on Vesuvius, without being able to get up the mountain. The Duke's lungs are very weak, and his body is ill formed and not strong, and the Duchess is rather corpulent. However, the novelty pleased them. The Duchess's feet came through her shoes, but I had luckily desired her to take a second pair.

For his pains, the Grand Duke Paul gave Sir William "a gold snuff box, all over with diamonds".

For the common of mortals, it was an arduous excursion, which was usually thought to require amiable companions, a servant or two, and a good supply of food and wine. The groups left Naples by coach and swept past the orchards, vineyards and the villages near the coast, which were becoming cluttered with labyrinths of new dwellings. They reached Resina, which had been built on top of Herculaneum. The adventurers had no sooner halted there than they would discern what they saw as a group of suspicious-looking rapacious brigands who emerged from the shadows to accost them, as if they had sprung from the depths of the ancient city itself. "Johnny Foreigner" was still a strange creature.

Then began the haggling and the bargaining, without which, it seems, no excursion to Vesuvius has ever been complete. In July 1867, Mark Twain, described their reception. "They crowd you – infest you – swarm about you and smell and sweat offensively, and look sneaking and mean, and obsequious . . . We got our mules and horses after an hour and a half of bargaining".

On 21 February 1845, Dickens recounted the almost operatic chaos:

233

There is a terrible uproar in the little stable-yard of Signor Salvatore, the recognized head-guide, with the gold band round his cap; and 30 under-guides, who are all scuffling and screaming at once, are preparing half a dozen saddled ponies, three litters, and some stout staves, for the journey. Every one of the 30 quarrels with the other 29 and frightens the 6 ponies.

The visitors had a choice of conveyance. They could be dragged up the mountain, like Goethe in 1787, clinging onto leather thongs attached to the belts of their guides; they could be carried in chairs or litters, as Shelley put it in 1818, "like a member of Parliament who has just won the elections"; or, like Prince Christian of Denmark and Sir Humphrey Davy in 1820, they could ride on mules or donkeys, and collect rock specimens and conduct small experiments on the way.

The groups passed through the vineyards producing the famous Lachryma Christi wines, which had long flourished on the lower slopes of Vesuvius, and which also attracted attention because the vines were trained over small poplar trees. Eventually, the Hermitage of San Salvatore came into view about half way up the western slopes of the mountain. It was a spot where everyone had to stop. The hermits who have lodged at San Salvatore must have been among the most sociable in Christendom. Scarcely a day seems to have passed without their welcoming visitors, who required less of a spiritual revival than a place to have a snack and rest their already weary limbs.

Old lavas

Many of the old lavas erupted since 1631 had entered the valley of the Atrio del Cavallo at the foot of the ridge of Monte Somma. They had then spread out onto a wide platform, near the Hermitage, that marked the lower southern sector of the rim of the old caldera of Somma volcano. So many flows have followed this path that they have now all but completely covered this rim – in marked contrast to the higher northern rim of the caldera, which still stands out as Monte Somma. The visitors soon realized with consternation that these lava flows were remarkably rugged, and that they had to be crossed. Their surface was a masochist's paradise. As Renato Fucini recorded in 1877, "stumbling at every step, we crossed over large fragments of sharp

The Hermitage of San Salvatore in the mid-eighteenth century, with Vesuvius and the Monte Somma ridge behind. At a height of 660 m, the Hermitage was at the half-way stage on the ascent of Vesuvius, where visitors could take refreshments. The building was more protected from eruptions than it appears to be: the northwestern rim of the caldera was high enough to divert lava flows, and the prevailing westerly winds usually blew airborne fragments away to the east. A site near by, with a similar good view of the cone, was chosen for the Vesuvian observatory, opened in 1845.

and lacerating lava that took the skin off our feet – and off our hands every time we had to make use of them to keep our balance".

The lavas inspired fear, phantoms, enchantment and, sometimes, even scientific analysis. On 18 December 1818, Shelley was enthralled. The flows seemed like "an actual image of the sea, changed into hard black stone by enchantment. The lines of the boiling flood seem to hang in the air, and it is difficult to believe that the billows which seem hurrying down on you are not actually in motion".

In July 1867, Mark Twain, who certainly had a way with words, was carried away on a verbal transport. These lava flows were:

. . a wild chaos of ruin, desolation, and barrenness – a wilderness of billowy upheavals, of furious whirlpools, of miniature mountains rent asunder – of gnarled and knotted, wrinkled and twisted masses of blackness that mimicked branching roots, great vines, trunks of trees, all interlaced and mingled together; all these weird shapes, all this turbulent panorama, all this stormy, far-stretching waste of blackness, with

235

its thrilling suggestiveness of life, of action, of boiling, surging, furious motion, was petrified! – all stricken dead and cold in the instant of its maddest rioting! – fettered, paralyzed, and left to glower at heaven in impotent rage for evermore!

What are phantoms for some are scientific challenges for others. In 1778, these lava flows stimulated the French traveller and diplomat, Vivant Denon, to remark how their appearance and vegetation cover changed with age, so that the relative ages of the flows could be distinguished. It was a technique that his fellow countrymen had developed in Auvergne, and one that was to remain a useful means of qualitatively dating surface lavas until archaeomagnetic studies developed:

> The most recent lavas show only ferruginous and bituminous cinders that are the colour of clinker. After a century, this calcination becomes less sharp and is first covered with moss or lichen, which deteriorates, and is then regenerated, collects and retains dust, then ferments, and soon broom and lavender begin to grow upon it. Then each type of vegetation comes in turn, and so on, until the next lavas arrive.

Molten lavas

Often enough, rather paradoxically, the visitors were less afraid of the effusive streams of molten lava than of the old solidified flows. The molten streams were fascinating rather than fearsome. They were extremely beautiful, especially at night, when the brilliant red and golden glow, silvery fumes, the heat and the sheer spectacle of their motion made the spellbound observers try all manner of ruses to get as close to them as they could. It was only on mature reflection afterwards that the power of nature caused them consternation. In 1790, for instance, the artist, Louise Vigée Le Brun, confessed that "the terrible lava . . . greatly troubles the soul; I could not talk about it when I returned to Naples".

On 10 June 1717, the philosopher, Berkeley, reached the "burning river" in the "horror and silence of the night". It was:

> . . . a scene the most uncommon and astonishing I ever saw. Imagine a vast torrent of liquid fire rolling from the top, down the side of the

236

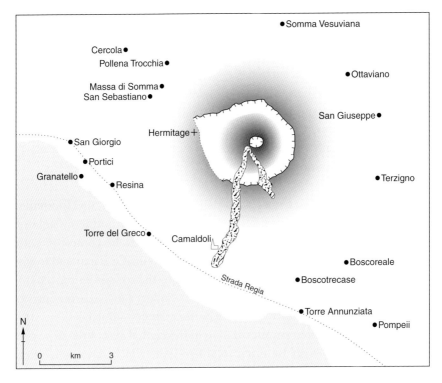

Lava flows emitted during the eruption in 1717.

mountain, and with irresistible fury bearing down and consuming vines, olives, fig-trees, houses; in a word, every thing that stood in its way. This mighty flood divided into different channels . . . the largest stream seemed half a mile [800 m] broad at least, and five miles [8 km] long . . . I walked . . . so far along the side of the river of fire, that I was obliged to retire in haste, the sulphurous stream having surprised me, and almost taken away my breath.

For those who were not completely overcome by the display, the molten lavas presented a chance to conduct some scientific enquiry. Goethe climbed up to Vesuvius on 20 March 1787 with his friend Tischbein. Half way up the cone, a vent was giving off turbulent clouds of steam, under which they could make out a lava flow. A closer look revealed how the flow of the molten stream soon confined itself to an axis within its sidewalls of solidifying lava:

The lava on both sides of the stream cools as it moves, forming a channel. The lava on its bottom also cools, so that this channel is constantly being raised. The stream keeps steadily throwing off to right and left the cinders floating on its surface. Gradually, two levels of considerable height are formed [like a pair of walls], between which the fiery stream continues to flow quietly like a mill brook. We walked along the foot of this embankment while the cinders kept steadily rolling down its sides. Occasionally there were gaps through which we could see the glowing mass from below.

On 26 January 1820, Prince Christian of Denmark and Sir Humphrey Davy also investigated a lava flow then issuing from the northern flanks of the cone. Prince Christian declared that it "could best be compared with molten iron as it is leaving a blast furnace; and as soon as it is exposed to the air, it is covered with a black crust coming from the oxygen in the air . . . Sir Humphrey Davy took the opportunity to conduct a few experiments on the lavas to demonstrate that they contain no carbonic substances."

Eventually the molten stream running in the central axis of the flow also cools and is roofed over by black solidified lava, which, however, often remains very hot for days. For instance, when Hans Andersen was on the mountain in March 1834, a new lava stream broke out just in front of his party. At length, they had to walk over the flow as it was solidifying:

> . . . only the crust had been stiffened by the air, and through the cracks in it we could see the fiery red beneath the surface; with our guide ahead of us, we stepped out on this expanse; it warmed us right through the soles of our shoes, and if the crust had given [way] we should have gone through into the fiery abyss . . . The smell of sulphur was extremely powerful, and the heat under our feet was almost too much to bear, and we only managed to stay there a few minutes.

Sometimes the admirers of the glowing lavas had to use their ingenuity. A calm interlude on 21 October during the eruption in 1767 encouraged the German art historian, J. J. Winckelmann, to climb up to the source of a hot new flow with two companions, three servants and a guide. Near the vent, they turned out to be more enthusiastic – and foolhardy – than their guide, who wanted to go back down the mountain and stay alive. The gentlemen brandished their sticks at him and forced him upwards. Then, when the lavas became too hot even for the foolhardy, they stripped off their clothes, skewered their snack of pigeons, roasted them over the glowing flow, had a picnic, and then retreated down the mountain, safe and sound.

The cone and its crater

Climbing the cone was the most arduous part of the whole trip. Some guides struggled to carry the ladies and older gentlemen on litters, others dragged or pushed up the fitter men, and the more athletic visitors among them made their own way to the top. But for every three steps they took upwards, they slipped back two more.

The cone of Vesuvius was composed of ash and cinders, which were especially loose on the surface and which sloped steeply at their natural angle of rest of about 32°. As John Evelyn remarked in 1645, the explosions took place every few minutes, "ejecting huge stones with an impetuous noise and roaring, like the report of many muskets discharging". Persistent activity had been mostly constructive, building up the cone with fragments and filling the crater with lava to the brim, although occasional more violent explosions had decapitated the cone and left behind a large fuming crater. The visitors who

Sketch, after an engraving by J. T. Smith, 1798, of the cone of ash and cinders, surrounded by lava flows, produced by persistent moderate eruptions of Strombolian type within the main crater of Vesuvius. Part of the wall of the main crater rises on the left.

reached the summit could witness at close quarters and in safety only the most moderate explosions, although they still made their hair stand on end. Thus, the descriptions of such events completely fail to do justice to the power of Vesuvius, for all their apparent hyperbole. Many of its more violent outbursts would simply have sent every spectator to oblivion: those who had been standing within several kilometres of the summit during an eruption as powerful as those in AD 79 or 1631 would not have survived a moment to tell the tale.

Observers sometimes saw how the moderate eruptions within the crater eventually filled it to the brim. In 1778, for instance, it was no more than 20 m deep after lava emissions had been filling it for a decade. Vivant Denon's party were able to walk into the crater and they were then unwise enough to cross hot lava that had erupted only the previous night:

> We strode over crevasses, from which . . . sulphurous fumes were issu-
> ing that were too hot to touch and gave off a suffocating vapour. The
> smoke from these mouths was making the fog even thicker and added
> to our fear at this terrible spot . . . we kept hearing the noise of explosions
> close by . . . We were walking on the crumbling crust, which looked
> like molten metal that had cooled. It was giving way under our feet all
> the time – not without adding to our fears . . . We reached a small hole,
> which, we were told, was one of the main mouths. One of the lads threw
> a stone down into it and it swished for a long time as it fell . . . [later]
> my leg suddenly sank into a crack in the lava that was covered with
> scorching sulphur. It made me realize . . . the dangers of approaching the
> mouths, whose edges could collapse into the abyss at any moment.

Most of the ash and cinders erupted within the main crater of Vesuvius piled up around the vent as a small cone. At times, the fragments were accompanied by emissions of fumes and minor lava flows. These little cones in the midst of the main crater provided a working model of how the whole volcano had been constructed, although it took many decades for this appar-ently simple truth to become accepted. As Dickens reported in 1845, "from a conical hill . . . great sheets of fire are streaming forth: reddening the night with flame, blackening it with smoke, and spotting it with red-hot stones and cinders, that fly up into the air like feathers, and fall down like lead".

Berkeley described his own impressions of the noises in the crater in his letter to the Royal Society:

On 17 April 1717, with much difficulty, I reached the top of Mount Vesuvius, in which I saw a vast aperture full of smoke . . . I heard within that horrid gulf certain odd sounds, which seemed to proceed from the belly of the mountain; a sort of murmuring, sighing, throbbing, churning, dashing (as it were) of waves, and, between whiles, a noise like that of thunder or cannon, which was constantly attended with a clattering like that of tiles falling from the tops of houses on the streets. Sometimes as the wind changed, the smoke grew thinner, discovering [revealing] a very ruddy flame, and the jaws of the pan or *crater* streaked with red and several shades of yellow . . . I could discern two furnaces almost contiguous, that on the left, seeming about three yards in diameter, glowed with red flame and threw up red-hot stones with a hideous noise, which as they fell back, caused the aforementioned clattering.

Berkeley returned to the summit on 8 May 1717 and "found a different face of things". The changes illustrated fairly typical developments that occurred within the crater during periods of persistent moderate activity. This time, the crater was "a mile in circumference and an hundred yards deep. A conical mount had been formed since my last visit . . . made up of the stones thrown up and fallen back again into the crater". The two furnaces were still there: that on the left now formed the main vent of the new cone and it "raged more violently than before, throwing up, every three or four minutes with a dreadful bellowing, a vast number of red-hot stones . . . at least 3000 feet higher than my head as I stood on the brink . . . The other mouth to the right was . . . filled with red-hot liquid matter, like that of a furnace of a glass[-blowing] house, which raged and wrought as the waves of the sea, causing a short abrupt noise like that which may be imagined to proceed from a sea of quicksilver dashing among uneven rocks. This stuff would sometimes spew over and run down . . . the conical hill; and appearing at first red-hot, it changed colour and hardened as it cooled."

Some visitors narrowly escaped the occasional stronger spasms. In March 1766, for instance, the future Earl of Bristol was among those who climbed up to inspect the crater, during what seemed to be an interval of volcanic repose. He was more than surprised when Vesuvius suddenly sprang back to life and narrowly failed to send him to eternity. A sizeable boulder hit him on the arm, tore his clothes, and made him bleed so much that he took a fever that confined him to his bed for five weeks.

Vesuvius versus Fucini, 1877
[*Napoli ad occhio nudo* 1878]

The Italian author Renato Fucini climbed Vesuvius on 27–28 May 1877. In the description he published in the following year, he displayed all his considerable literary talents in an explosion of imaginative impressions that was the very opposite of any laconic scientific account. On first seeing the solid old lavas of yesteryear, he exclaimed:

> My God! What desolation. What an impressive silence . . . I imagined that I was among the blackened remains of a battle of giants. I looked around in terror. It seemed as if the limbs of the giants, some smashed, and others intact, were bursting out from an enormous mass of ruins. Torsos, thighs, and arms seemed to have been spread higgledy-piggledy in a vast field of death. I saw gigantic reptiles . . . the flanks and hind-quarters of horses and monstrous animals that had been broken and scattered about. I saw the remnants of tents, lacerated and burnt clothing, gun carriages, bombs, mortars and destroyed forts and piles of ropes.

He climbed on up the cone with a mixture of trepidation, curiosity and uncontrollable elation:

> The slope is very steep and . . . you sink into it up to your knees, so that when you have taken ten steps at the cost of unparalleled fatigue, you have hardly moved one metre forward . . . I really began to ask myself if it wasn't dangerous to go on. Would I be able to breathe enclosed in that mantle of sulphurous vapour? . . . I honestly admit that I was near to fear and panic. I had a nasty moment when the wind got up a little, and I saw the enormous bulk of the plume suddenly collapse, pour quickly down the flanks of the mountain, and rush towards me, turning and swirling across the steep, bare slopes . . . I was completely overpowered within seconds. My eyes started to water. I was sneezing. I had some bouts of coughing . . . I started to shout, and sing, and call out to my friends, although I could no longer see them through the thick fog . . .
> [At the summit] our enthusiasm turned into frenzy. Everybody was talking at the same time. There were cries of amazement; handshakes all round; glasses of wine were raised; men ran hither and thither . . . I shouted greetings to my parents, to my friends, and even to my enemies . . . I would have embraced Lucifer himself if he had come up to mock us in his mantle of flames.

The crater exerted its habitual fascination, drawing Fucini and his friends like bees to a honey pot. They just had to go down into the fuming hole, and the guide eventually agreed to go with them. But first, they downed another glass of wine.

> The whole spectacle vaguely resembled a furious hurricane. Imagine the noise

of wind swishing through a forest of fir trees; the noise of electricity discharging underground; the blows of a huge hammer beating down onto a vast sheet of copper . . .

Our hands and feet were burnt . . . We were half-blinded by the blasts of acid, boiling steam, and confused and sweating. We could see the infernal crucible boiling a few metres below us. It was absolutely impossible to go any farther down . . . I would have needed more than a thousand eyes to take in all that sombre beauty at a glance. I felt so small . . . My companions [seemed] but fantastic shadows seen through a feverish dream . . . We were not afraid and had no apprehension of danger; we were drunk with the roaring and the fire; and if a burst of lava had covered us, we would have died crying out with joy like a madman who sees that the clothing on his back is on fire . . .

The most terrifying point was the mouth of the eruption that we could see a few metres below us. The formidable noises that arose from the bottom seemed like a pride of howling lions resounding from the flames of the monstrous furnace. In the throat itself, the fluid [molten lava] was stirring continually, rising and falling swiftly in a swirling mass, and boiling and groaning dully until it bulged in the centre, and gradually swelled up into great blisters. Eventually, the blisters would burst with a violent explosion and swirls of fiery fumes and clots of lava like blood-covered rags were thrown into the air almost as high as where we were standing. Some were still liquid, but others had already partly cooled down. Then they fell back into the crucible with a sinister crackling like a rain of stones . . . With each explosion came a flash like an electric discharge that was reflected brilliantly on the plume and its fiery base . . . It seemed as if I had smoked opium or taken hashish . . . I seemed to be in a feverish dream . . . I do not believe that there is a more sublime spectacle . . .

Renato Fucini encounters the lava fields of Vesuvius. Similar lavas pictured above are in the Lava Fields National Monument, northern California.

243

Descent

After such spectacular displays, the descent from the crater would have been an anti-climax had it not been so rapid. Most of the parties strode manfully down slope, up to their knees in ash, and they reached the base of the cone in a few minutes. As Hans Andersen put it, "we had continually to dig in our heels to prevent ourselves from tumbling down on our faces; we preferred to fall on our backs in the soft ashes. The descent was a case of dropping merrily through the air". Larger parties, such as that including Charles Dickens, suffered from a wider range of problems:

> Ten or a dozen of the guides cautiously join hands, and make a chain of men; of whom the foremost beat, as well as they can, a rough track with their sticks, down which we prepare to follow. The way being fearfully steep, and none of the party . . . being able to keep their feet for six paces together, the ladies are taken out of their litters, and placed, each between two careful persons; while others of the 30 [guides] hold them by their skirts, to prevent their falling forward.

Eventually, they reached the Hermitage, where they enjoyed "a cheerful meal", had a good rest before a blazing fire, then took horse again, and slowly made their way back to Resina. Life returned to normal almost as soon as the adventurers set off for Naples again: balls, receptions, dinner parties – back to civilization. Just like home.

Further reading

Berkeley 1717; Denon 1997; Dickens 1846; Fucini 2004; Goethe 1816–1817.

Chapter 11

Persistent activity 1822–1944: scientific scrutiny

Persistent eruptions of Vesuvius encouraged frequent scientific analysis of the volcano, and led to the establishment of a permanent observatory on its flanks.

During the nineteenth and twentieth centuries, the persistent pulsations of Vesuvius continued unabated. There was an exceptionally long period of persistent constructive activity between 1872 and 1906, when the main cone reached its greatest height recorded in modern times (1335 m). On the other hand, Vesuvius also unleashed four powerful "final eruptions" in 1822, 1872, 1906 and 1944, which badly damaged its uppermost parts. The last three of these outbursts also caused losses of human life.

It was during the nineteenth and twentieth centuries, too, that observation of the behaviour of the volcano changed from mainly anecdotal descriptions to more analytical studies of an increasingly scientific nature. The Vesuvian observatory was established and instruments recorded features such as earthquakes and gas emissions, which warned of eruptions to come. Surveillance of the volcano made its tentative beginnings. Vesuvius trained its students too, as, for example, Monticelli, Scrope, Palmieri, Mercalli, Perret, and Imbò made their names as some of the most accomplished volcanologists in the scientific world, as well as in the Neapolitan press.

The eruption of 1822

For many centuries, there was a common belief among those who lived near volcanoes that they did not erupt during good weather. However, events in 1822 gave the lie to that flawed notion. Brilliant sunshine had made 20 October one of the most glorious days of that Campanian autumn, but early in the evening Vesuvius began its most spectacular outburst of the nineteenth century. The volcano spewed out so much ash, cinders, rocks and lava that many trembling observers were sure that they were about to succumb to a repetition of the cataclysm that had buried Pompeii. Luckily, their fears proved groundless because this powerful "final eruption" was at least a whole order of magnitude less than that in AD 79. Paradoxically, the eruption of 1822 demonstrated that the most powerful eruptions do not always cause the greatest destruction. In particular, no lava flows invaded highly populated zones, no towns were overwhelmed, and no lives were lost. Vesuvius itself was the chief casualty: it lost the top of its cone.

In 1818–1819, the British politician and volcanologist, G. J. Poulett Scrope, visited Vesuvius, when he was making his first forays into European volcanic regions. He did not realize that Vesuvius was preparing its next major effort, because the crater was already full to the brim, "covered with blocks of lava and [cinders] and cut up by numerous fissures, from many of which clouds of vapours were evolved".

On 20 October, the ground quaked several times, and everyone turned apprehensively towards Vesuvius. It was only then that many people recalled that the wells, and even the streams, had recently been drying up around the flanks of the volcano, and several earthquakes had shaken the area during the past week. Some had even heard odd and inexplicable loud bangs. Some also remembered that Vesuvius had ended its previous eruption on 28 February that year without expelling the usual cloud of white ash that would normally have signified the conclusion of its eruptive efforts. Thus, led by the Neapolitan scientist, Teodoro Monticelli, more informed observers had every reason to fear another outburst. They did not have long to wait.

After nightfall, noisy explosions were shaking the ground, glowing ash was rising up through the clouds of black fumes, and lightning was flashing all over the mountain. At mid-day on 21 October, lava gushed out of the western flanks of the cone and piled up in the Atrio del Cavallo. In the early hours of 22 October, lava fountains spurted some 600 m skywards from the crater, smashed through its eastern rim, and surged 2 km into the air in a great column of fire that lit up the apprehension on the faces of spectators in Naples itself. Fumes and glowing fragments rained down for three-quarters of an hour onto the southern slopes of the volcano, where they gave off such a disgusting smell that many people retreated to Naples; these refugees only added to the confusion in the capital. Next day, the Neapolitan newspaper, the *Giornale del Regno delle due Sicilie*, clearly keen on law and order, remarked that "the police took all the necessary steps to avoid all the inconvenience that might have arisen".

By dawn on 22 October, the volcano was "roaring like a tempest at sea"; ash went on falling, and experienced men with initiative and forethought began to shovel its thickening layers from their rooftops to stop them from collapsing upon their families. Yet more lava began to flow from the cone and divided into three branches directed towards Torre del Greco, Resina, and the Hermitage of San Salvatore.

At 1 p.m. on 22 October, a fearsome explosion developed a Plinian column that rose 3 km above the volcano. As Monticelli and Covelli later recounted, the trunk was a reddish-tawny colour, the billowing parts shone in the Sun like driven snow or lumps of the whitest cotton, and the shaded northern parts were a fine turquoise blue. Its allure did nothing to reduce its menace. At 4 p.m., the eastern rim of the crater collapsed, a great volume of lava issued forth, streamed down hill, and piled up on the outskirts of Boscotrecase in a threatening tongue more than 13 m thick.

During the evening of 22 October, strong earthquakes warned of yet more activity to come. At 8 p.m., another great fountain of lava spurted skywards. A great fissure opened and exploded ash from five vents above Torre del Greco. Then at 1 a.m. on 23 October, with a loud roar and a violent earthquake, the walls of the main crater caved in, and a vast cloud of ash and fumes shot from the cone. The ash fell unusually heavily, and all the settlements on the southern flanks of Vesuvius seemed in imminent danger. In Boscotrecase, the church of Santa Anna collapsed under the weight of the ash, and many roofs gave way in Torre Annunziata.

Meanwhile, more lava was spilling down the lower slopes of the volcano. One stream blocked part of the Strada Regia between Torre del Greco and Torre Annunziata, another threatening mound of lava accumulated above Boscotrecase, and yet another destroyed the Prince of Ottaviano's forest, although this flow halted before it could consume his country cottage. By 3 a.m. on 23 October, the supply of lava, ash and cinders came to a halt and induced false hopes that the crisis might be over.

At 2 p.m. on 23 October, activity resumed with the usual fanfare of explosions and earthquakes. Vesuvius expelled a pair of columns. Just before dawn on 24 October reddish dust emerged and spread out mainly northwards. Gloom prevailed all over the volcano and even in Naples itself.

On 24 October 1822, the *Giornale del Regno delle due Sicilie* reported that:

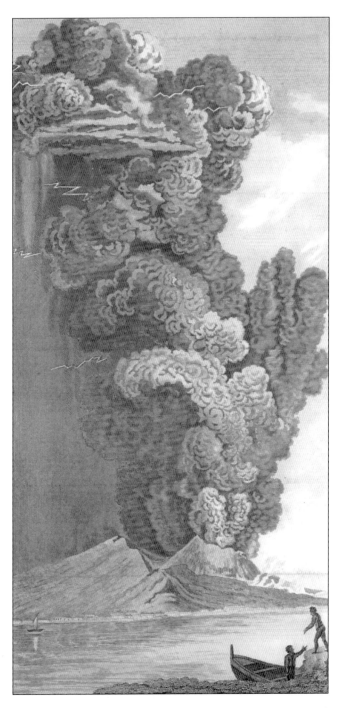

The towering eruptive column of the eruption of 1822 (Scrope 1825).

Monticelli and Covelli describe some reactions to the eruption

Fleeing people blocked the main road from Torre to Resina and the road from Ottaviano to Naples. Terror had spread throughout the regions around Vesuvius . . . The frequent earth tremors, the uninterrupted rain of red-hot stones; the continual explosions of thunderbolts hurled violently onto the highest points of the churches, the houses and the trees, the widespread flashes of lightning that sprang as often from the ground as from the sky – all produced different effects upon the unfortunate people caught in their homes . . . The river of fire surged out . . . [but] . . . the terrified inhabitants of Boscotrecase did not apprehend their imminent danger until the current reached within a mile of their settlement . . . The more the wretched people tried to dash from their threatened homes, the more were they kept back by the sand [ash], by the glowing stones, or by the lightning that fell upon them at every step. Not only did the fiery rain cover the ground with stones, but great clumps of fire [clots of lava] were still falling from the air and causing great damage when they crashed down onto the buildings. The most terrified inhabitants then tried one final means of reaching safety. Some of the bravest among them covered their heads with pillows or tables and left their homes . . . for places where they believed they would be in less danger. Those who stayed at home, were afraid that their roofs and walls would collapse. They were more discouraged by the lightning than by the falling sand, and took shelter under the arches of their own houses . . . The desolation was widespread from Ottaviano, in the north, to Torre del Greco, in the south. Not only was the air filled with the cries of the distraught people, the rumbles of the explosions, and the thunder, but also by howls of the domestic animals, who had been shut in their stalls and were struggling violently to break the cords that fastened them and make their escape to safety.

About 800 people from neighbouring hamlets were welcomed into the prefecture [of Naples] by order of the police, where they received both financial aid and all that charity demands in such unfortunate conditions. Other grants were given to neighbouring places [where] the people had congregated after fleeing from the force of the rain of ash, which, more than anything, had destroyed the lands on the flanks of the mountain.

By mid-day on 25 October, the erupting column had reached majestic proportions. The wind pushed its turquoise-blue trunk to the north over Somma and Ottaviano, so that it seemed to be leaning over. The *Giornale* reported that "the number of people fleeing from Somma and Ottaviano has greatly increased since yesterday. Last night 2000 individuals were lodged in the prefecture, where they received financial aid for their sustenance".

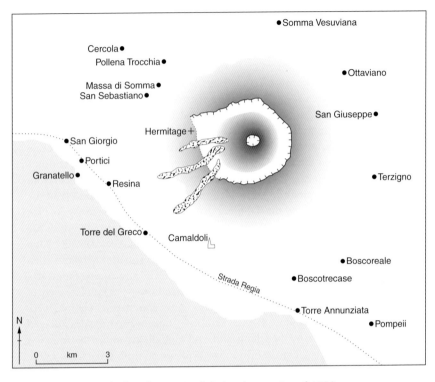

The lava flows emitted during the eruption of 1822.

There seems to have been unrest in Naples, caused initially, no doubt, by the fearsome sight of the volcano in fury, although it would almost certainly have been increased by the influx of terrified and ragged refugees from the countryside. However, the problems have to be inferred from the *Giornale* on 28 October, where the loyal newspaper praised the administration and claimed that the firmness, zeal and actions of the civil authorities had been matched only by those of the army, who had organized frequent patrols to keep good order in the countryside. What action had been taken, and against whom, was not revealed. Meanwhile, the two local infantry regiments had not only been trying to keep the roads clear of ash and cinders. They had also taken the sensible precaution of loading the powder from their depot onto vessels and transporting it out to sea. In this way they were able to prevent the hot ash from igniting the gunpowder and thereby blowing the depot and its surroundings, not to mention the infantry, to smithereens.

At last, at 10 a.m. on 2 November, the most welcome portent of the eruption appeared. Vesuvius gave out the white ash that promised an end to the outburst. Nevertheless, these concluding eruptions themselves lasted for an exceptionally long time and did not cease until 10 November.

The white ash marked the end of the eruption, but not the end of the misfortunes of those living on the flanks of the volcano. Vesuvius was mantled with a cover of loose dust and ash that ranged from 12 cm thick in Torre del Greco, to 24 cm in Ottaviano, and at least 1.5 m on the cone itself. This material was unconsolidated, thoroughly soaked, and lying precariously on steep slopes; it could therefore easily be set in motion by the rains. It had been raining for the best part of a week, and the ash was already choking the usual stream channels that could have evacuated the rainwater efficiently. All the ingredients that could generate large mudflows were therefore assembled, and widespread mudflows duly developed. Elsewhere, however, the rain transformed the loose surface debris into a layer like concrete. Here, the rainwater could not sink into the ground, but flooded down hill after every storm. The floods and the mudflows had much the same destructive effects, and often could scarcely be distinguished, although if anything the mudflows caused more damage. As G. J. Poulett Scrope wrote:

> Torrents of sand, mixed with water, appearing like liquid mud, swept, with terrible impetuosity, down the slopes, tearing them up in their passage, hurrying along fragments and blocks of lava, of great size (some even of about 15 m in girth), and depositing heaps of alluvium on the sides and at the foot of the mountain. The damage occasioned by these *lave d'acqua* [floods] or *di fango* [mudflows], as they are called in the language of the country, was far greater than what was suffered from the *lave di fuoco* [lava flows].

The floods and mudflows swept down the slopes of Vesuvius after every rainstorm until well into January 1823. It seemed as if the whole area would be swamped, and many concluded that only an immediate evacuation would save them. Little palliative action could be taken against the mudflows at the time, but the impenetrable concrete-like layer could at least be broken up. Instead of bewailing their fate in the churches, gangs of able-bodied men made use of the experience gained from previous crises, climbed to the higher slopes of the mountain, and smashed the surfaces of the solid layers so that the water could sink into the ground. The citizens of Ottaviano, Boscotrecase and Resina were notably quick to act, and these towns suffered much less destruction than many others.

Locations of eruptions of Vesuvius and casualties

The eruptions of Vesuvius have a strong central focus. Most of the events take place from the crater of the central cone and comprise both effusions of lava and explosions of fragments. Some eruptions take place from about half way up the flanks of the central cone, and these too can be both effusive and explosive, but are commonly less explosive than those from the central crater. A few eruptions are eccentric, in that they take place beyond the main cone and occur along two sets of fissures transecting the old Somma volcano: one runs from north to south through the base of the volcano; the other trends from northeast to southwest. Although these eruptions are not particularly violent, they often occur at relatively low altitudes, making nearby settlements vulnerable to damage, as happened in 1779, 1794 and 1861.

The greatest outbursts in historical time, in AD 79 and 1631 (and possibly the eruption in c. AD 472) caused by far the largest numbers of fatalities. Many thousands must have died in AD 79 and at least 4000 were killed in 1631 – ten times the total number of victims claimed by all subsequent eruptions. After 1631, there was no direct relationship between the violence of the eruption and the number of casualties. Although the deaths that occurred near the cone in 1872 were an exception, the number of fatalities (and the damage) usually depended on the position of the settlements in relation to the paths taken by the most destructive materials erupted, especially if the vents were situated low on the flanks of the volcano near vulnerable centres.

On 23 November 1822, Vesuvius was visited by no less distinguished a pair than the great Earth scientist, Alexander von Humboldt, and his sovereign, the King of Prussia. The "final eruption" was almost spent, but its dying throes within the crater were still impressive. To the visitors, the crater looked like the gates of Hell, giving off multitudes of fumes, oppressive heat, and an intolerable smell. The open fissures on the sidewalls were exhaling both hot fumes and lava, and every now and again loosened boulders crashed, echoing into the abyss, where molten lava was still bubbling, boiling, stirring and turning in turmoil in its enormous crucible.

By the time the eruption ended, about $165\,000\,m^2$ of woodland and over $100\,000\,m^2$ of vineyards had been lost. The Neapolitan government took the appropriate action to recompense the victims. The eruption had decapitated the volcano once again, lowering the summit by about 93 m. At its crest lay a gaping crater, 216 m deep.

The eruption of 1872

By the middle of the nineteenth century, Vesuvius was famous not only among upper- and middle-class tourists, but also among professional and amateur scientists. Sir Charles Lyell and Alexander von Humboldt, for instance, had made the Earth sciences popular, more and more geologists were studying volcanoes, and Vesuvius, too, seemed to be increasingly agitated as the years went by. The vigorous Neapolitan scientific community was also stimulated to develop an even greater interest in its volcano. Thus, in 1841, the physicist Macedonio Melloni proposed to King Ferdinand II that an observatory should be established on the slopes of the volcano, to foster volcanological studies in general and to keep an eye on Vesuvius in particular. Thus, the world's first volcanological observatory was built and opened in 1845, and Melloni became its first director. In 1855, Luigi Palmieri was appointed director and he installed his own seismograph and other equipment that he had bought in Paris. He was an ideal person to launch the scientific study of the volcano, for he was so obsessed with the whims of

Sketch of the eruption of 1872, based on a photograph taken by Giorgio Sommer in the afternoon of 26 April. The stratospheric winds are blowing the crest of the eruptive Plinian column to the right (towards the northeast). The mid-altitude winds are blowing in the same direction, but the fumes emerging from the lava flow advancing towards San Sebastiano, on the lower left, are being blown to the left (towards the southwest).

Vesuvius that he rarely left his post. Some of Palmieri's authoritative studies concerned the eruption in 1872.

During 1871, Vesuvius had erupted repeatedly, both from the central crater and from a new fissure on the northern flank of the cone. The last eruptive episode of the year ended in November 1871. After a Christmas recess, both vents resumed their activity in about the middle of January 1872, with moderate explosions of cinders, emissions of lava clots and small lava flows that continued during February and March. More serious events took place at the beginning of April 1872, when lava poured over the rim of the main crater and ran down the north side of the main cone. On 8 April, the fissure that had formed on the northern side of the cone in 1871 opened once again, lengthened, and spurted out lava, ash and steam.

The vigour of the eruptions increased after 24 April when powerful explosions gave off great clouds of reddish-yellow and blue steam. The seismographs in the observatory became quite agitated. At about 4 p.m., a mass of lava gushed from the main crater and reached the foot of the cone in a couple of hours. At 7 p.m. on 24 April, the new fissure itself started to expel great masses of lava. In no time at all, the whole sector of the cone facing Naples seemed to be ablaze from top to bottom; this spectacle lasted most of the night. There was something of a pause on 25 April, although the main crater continued to erupt apace.

The glorious sight of Vesuvius yet again "in flames" encouraged many tourists to climb up the mountain for a closer view. Many of the most inquisitive, foolish or courageous among them, accompanied by inexperienced guides, clambered over the extremely rough surface of the flows that had erupted in 1871. Although they had been warned of the dangers, one group entered the hollow of Atrio del Cavallo, lying between the main cone and Monte Somma, and went up to take a close look at the small eruptions coming from the vents at the foot of the main cone. From there they enjoyed the added bonus of a grandstand view of the booming explosions that were sending great splays of glowing fragments soaring from the main crater. A great eruption did not seem imminent and there appeared to be no pressing reason to suppose that a disaster was nigh. Indeed, the diligent Luigi Palmieri had, for once, left his post at 7 p.m. and gone to Naples, leaving his assistant, the priest Don Diego Franco, in charge.

At 3.30 a.m. on 26 April, a loud explosion shook the observatory to its foundations and awakened Don Diego. It seemed to him as if a great steam boiler in the middle of Vesuvius had blown up and had blasted out a huge notch on the northwestern sector of the cone. The fissure that had developed on 8 April now split open right under the feet of the hapless bystanders in the Atrio del Cavallo. Their screams echoed around the mountain.

As Diego Franco later recounted, part of the group was crushed by the fall of the projectiles, others were burned and asphyxiated by steam, ash and acidic vapours. They included eight young medical students. The exact number of fatalities was unknown, although 13 or 14 victims were transported to the observatory, where Don Diego and a visiting priest gave them the last rites before they died.

In 1887, Luigi Palmieri reported a slightly different version of events:

A large fissure appeared on the northwestern flanks of the main cone, and an enormous and vigorous torrent of fire surged out from its lower reaches. At the same time, the main crater gave off blazing projectiles that fell in broad arcs into the Atrio del Cavallo, which was already filled with fumes from the increasing eruption. The unfortunate men and women were either immersed in the fumes, surprised by the lava or stricken by the burning projectiles. Some were buried beneath the igneous torrent and disappeared. The dead bodies of two victims and eleven seriously injured people were recovered. One of the injured, Dr Antonio Giannone, a lecturer at the university, died in the observatory, and the remaining ten died in hospital.

There is a curious laxity about these two reports. It is not clear how many people were killed, where they died, nor exactly what killed them. The victims could have been killed by projectiles, or asphyxiated, or burned by steam, hot ash or acid fumes. Such a lethal combination might indeed have come from a violent gas explosion, or even possibly from a small pyroclastic flow. It is also probable that the gushing lava flow buried some of their bodies. The panic was apparently so great that no-one seems to have been able to find, or count, all the victims. The general consensus indicates that at least 12 people met their deaths in that awful instant. A memorial was erected to them at the observatory and they became part of the folklore of Vesuvius. For instance, during Renato Fucini's ascent of the volcano in 1877, he visited the spot where they had died, and typically added to the pathos by claiming, rightly or wrongly, that one couple had just been married and that another pair had just become engaged.

The notoriety of this tragedy spread, partly because it was so unusual. At least as far back as 1631, Vesuvius had never claimed so many victims so close to the main cone. There were two reasons for this apparent paradox. Most of the larger eruptions had come from the main cone and had expelled ash and pumice that caused roofs to collapse and result in fatalities several kilometres away. Visitors did not generally venture to the rim of the crater in such conditions, although there had been some lucky escapes. Eruptions taking place from vents situated low down on the flanks of the mountain have been the second main cause of deaths, such as those in Torre del Greco in 1794. However, such eruptions are very rare on Vesuvius and they commonly gave off lava flows that all but the sick and the physically handicapped could avoid with relative ease. Thus, the victims in 1872 were doubly unlucky when a flank fissure seems to have opened under their very feet at the same time as the main crater was showering them with fumes and red-hot fragments. Sir William Hamilton had himself narrowly escaped from a flank eruption in a similar spot on 19 October 1767 and he had not stopped running for nearly 5 km. In 1872, the victims had little chance even to set off.

Although Vesuvius had claimed its victims, it was far from spent. At about 7 a.m. on 26 April, the climax of the eruption began, with a salvo of fumes and ash that rose in the shape of a great pine tree high above the volcano. Then the first of several new vents opened in the Atrio del Cavallo. Almost everyone fled from the observatory; only Don Diego Franco remained, with the support of the concierge and a manservant. Then a great flood of lava emerged, streamed down the valley below the observatory and surged on towards Resina. The ground kept on shaking. At 9 a.m. another vent opened on the southwestern side of the cone, and lava began to flow towards Torre del Greco and around the hill where the Camaldoli monastery stands. An hour later, another great cannonade heralded the opening of even more vents in the Atrio del Cavallo. The lava swept out in a thick stream towards the villages of Massa di Somma and San Sebastiano. Palmieri returned to the observatory and spent the night of 26–27 April at his post. The volcano roared throughout the night. To Palmieri, it seemed as if the cone was "was completely perforated . . . as if it were sweating out fire". The ground shook continually. The lavas had surrounded the observatory, but Palmieri was undeterred, remarking that "The spectacle of this noisy eruption was truly sublime". The observatory and its crew escaped shaken but unscathed, because the site of the building had been carefully selected, high above the adjacent valleys, which made it safe from all but a lava deluge, or perhaps a pyroclastic flow, the like of which had not been witnessed for at least 240 years.

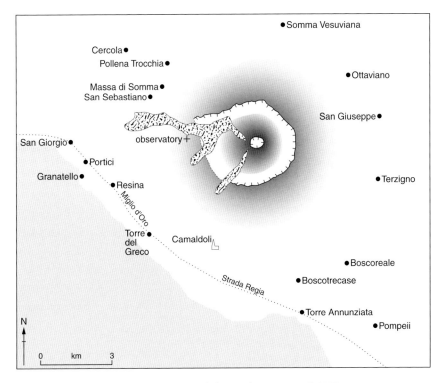

The lava flows emitted during the eruption of 1872.

Meanwhile, on the northwestern slopes of the volcano, the lavas that had issued from the Atrio del Cavallo on 26 April had continued on their journey and swamped parts of Massa di Somma and San Sebastiano (both of which had also been badly damaged in 1855). Massa di Somma lost about one third of its houses, and San Sebastiano almost a quarter. No-one was killed, because the flows had advanced so slowly that the villagers had time to collect their possessions before they ran away. The lava devastated good agricultural land and defoliated the trees and shrubs, so that it looked as if winter had suddenly returned to the spring landscape. However, the flow running nearest to the observatory produced an unusual and unexpected feature that intrigued Luigi Palmieri. Three times it gave off great balls of black smoke and clots of lava in the course of unusual explosions that lasted some 20 minutes each. It seems that they were caused when masses of gas exploded from the molten lava.

On 28 April, the lavas stopped flowing, only to be replaced by a succession of explosions that sent clouds of ash and pumice many kilometres from the main crater. Lightning played about the summit; a powerful thunderstorm

257

Major eruptions of Vesuvius, 1796–1944

1796 15 Jan. – 13 Oct. 1805	Persistent Strombolian activity (S)
1805 14 Oct. – 19 Oct.	Lava fountains with flow to the sea and to outskirts of Torre del Greco (I) (VEI 2–3)
1806 27 Jan. - 20 Oct. 1822	Episodes of persistent Strombolian activity (S), with an (I) over Christmas in 1817
1822 21 Oct. – 16 Nov.	Lava flows towards Torre del Greco and Boscotrecase; strong explosions; mudflows; most powerful eruption of the nineteenth century (F) (VEI 3+)
1834 23 Aug. – 2 Sept.	Collapse of the crater with flows towards Poggiomarino Destruction of Caposecchi and San Giovanni (F) (VEI 3)
1835 13 March – 31 Dec. 1838	Persistent Strombolian activity (S)
1839 1–5 Jan.	Eruptions from a fissure and large lava fountains (F) (VEI 2–3)
1841 20 Sept. – 4 Feb. 1850	Persistent Strombolian activity (S)
1850 5 – 16 Feb.	Lava flows from fissured N side of cone, and powerful explosions (F)
1855 1 – 27 May	Lava flows from fissure on N side of cone towards San Sebastiano, Cercola and Massa (F) (VEI 2–3)
1855 19 Dec. – 25 May 1861	Persistent Strombolian activity (S)
1861 8–31 Dec.	Lavas issued from a low-lying fissure near Torre del Greco, which damaged the town; eruption ended with ash emissions (F) (VEI 3)
1868 15–26 Nov.	Lava flows from fissure on NW flanks of cone, and large eruptive column developed (F) (VEI 2–3)
1871 13 Jan. – 5 Nov.	Persistent Strombolian activity (S)
1872 24–30 April	Powerful eruption with lava fountains and flows emitted from a fissure on the NW flanks of the cone; flows overwhelmed parts of San Sebastiano and Massa; new crater was 250 m deep; at least 13 deaths (F) (VEI 3)
1875 16 Dec. – 3 April 1906	Almost continual persistent Strombolian activity, notably in 1881–1884, 1891–1894, 1895–1899 and 1903–1904 (S)
1891 7 June – 3 June 1894	Strong emissions of lava and formation of the dome of Colle Margherita (I)
1895 3 July – 7 Sept. 1899	Strong emissions of lava; formation of the dome of Colle Umberto (I)
1906 4–22 April	Powerful eruption; lava flows destroyed part of Boscotrecase; ash and rock falls damaged Ottaviano; 218–227 deaths; the most powerful eruption of the twentieth century
1913 5 July – 2 June 1929	Persistent Strombolian activity fills the crater (S)
1929 3 – 8 June	Strong explosive eruption with lava fountains 500 m high and lava flows to the eastern base of the cone (I)
1929 13 July – 17 March 1944	Persistent Strombolian activity (S)
1944 18 March – 4 April	Powerful eruption with lava flows that damaged parts of Massa and San Sebastiano; lava fountains more than 2 km in height; and some small pyroclastic flows; 28 (possibly 47) deaths (F) (VEI 3)

VEI = volcanic explosivity index (devised to compare the power of different eruptions)   S = persistent moderate effusive–explosive Strombolian activity   I = intermediate eruption   F = final eruption

N.B. The types of eruption are not as clearly distinct as this table implies.

broke out to add its voice to the cacophony. However, this time the floods that often marked the end of Vesuvian eruptions did not take place. Indeed, all these alarming features had calmed down when the volcano fell dormant on 1 May, although for several days afterwards the whole area on and around the volcano was clothed in newly erupted white ash.

It was said that, although some citizens had left Naples during the eruption, many others had stayed in their homes because they knew that Luigi Palmieri was at his post at the observatory and would therefore look after them – he had become a sort of modern-day San Gennaro.

Agitation 1875–1906

All remained calm on Vesuvius for three and a half years until moderate effusive–explosive eruptions began deep inside the crater on 18 December 1875. From that date until 1906, the behaviour of Vesuvius changed a little in character. Its eruptions became longer, more frequent and often more agitated, and the volcano gave off volumes of lava that piled up both within the crater and in the Atrio del Cavallo. On the other hand, Vesuvius delivered none of the awesome destructive explosions that it had so often displayed before. As a result, these three decades were probably among the most constructive that Vesuvius had experienced for almost a thousand years. At the same time, the volcano added to its repertoire what seems to have been a rare feature: many rather viscous flows erupted from two spots beyond the base of the main cone. They were too viscous to spread far from their source, but piled up instead around the vent and formed a couple of lava domes. The dome of the Colle Margherita, named after the Queen of Italy, erupted between 1891 and 1894, and the Colle Umberto, named after the king, followed suit on the western base of the cone between 1895 and 1899.

The eruption of the Colle Margherita began on 7 June 1891 when a fissure opened on the northern flanks of the main cone and began to exhale fumes and steam. Soon afterwards, lava issued from the floor of the Atrio del Cavallo, and nearly overwhelmed the British geologist H. J. Johnston-Lavis and his assistant. They had to scramble up the face of the steep southern face of the Somma ridge, but for the rest of the night they were rewarded for their pains "with one of the finest spectacles one could be present at with any chance of safety". These short viscous flows continued to accumulate on the Colle Margherita until 5 February 1894.

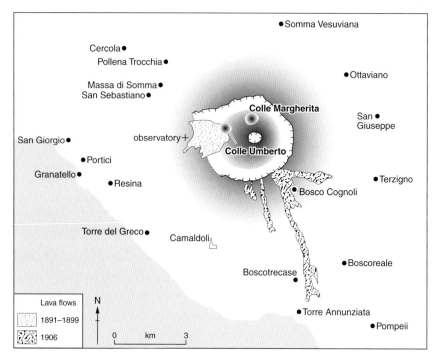

The lava flows emitted during the eruptions of 1891–1899 and 1906.

The eruptions that gave rise to the second dome, the Colle Umberto, began on 3 July 1895, when a fissure opened and gave off lava near the north-western base of the cone, near the observatory. These eruptions continued until September 1899, when the Colle Umberto was about 1 km in diameter and rose 160 m above its base, which itself lay about 740 m above sea level. Giuseppe Mercalli noted every detail of its growth and thereby confirmed his reputation as an eminent volcanologist, which he was to enhance with his studies of the "final eruption" of Vesuvius in 1906.

Meanwhile, the central crater at the crest of the main cone had been idle only for short periods. Most of these eruptions of ash and lava flows raised the level of the lava higher and higher within the crater and built up a small cone in its midst. Eventually, the ash and lava completely filled the crater, and the crest of the little cone could be seen from Naples rising above its rim. Thus, the mountain gradually increased in altitude until, on 22 May 1905, Vesuvius reached a height of 1335 m, the maximum recorded in modern times. Then, on 27 May, two vents opened in quick succession at the north-western base of the main cone and both started to emit lava flows. Within

Alfred Lacroix observes a lava flow at its source in 1905

On the evening of 3 October 1905, Alfred Lacroix, the eminent French volcanologist, who had just completed his classic work on the eruption of Montagne
Pelée in the West Indies in 1902, was taken to the main cone by Raffaele Matteucci
(director of the observatory), where lava was issuing from a small vent in a typical
effusive eruption:

> The opening was a little over a square metre in area; the incandescent magma
> was coming out quickly, at a speed of 6 m a minute . . . and ran straight down
> the steep slopes of the cone. For about the first 25 m . . . its fluidity was such
> that a stick could easily be plunged into the moving lava. But large stones
> thrown onto the moving mass did not sink; they welded quite firmly to the
> surface, stayed there, and were carried along by the current, in spite of the
> steepness of the slope. After the first 25 m, solidified cinders began to appear
> on the surface and they were immediately pushed to the edges, where they
> formed a kind of moraine, which quickly broadened. Thus, the cinder-free
> central part of the flow became narrower and narrower until it was com-
> pletely covered by the time it had travelled 100 m from its point of departure,
> and its surface was uniformly covered with a trail of smoking and incandes-
> cent cinders. The lava emerged in a continuous fashion, without projections,
> and without effort . . . And it spread out towards the Colle Umberto in a lake
> of fire lighting up the darkness of the night.

the next ten months, lava piled up, 50 m thick, between the Colle Umberto
and the base of the main cone. More explosive activity then began to throw
out ash and cinders during February and March 1906.

The nearly constant activity of the volcano and improved communications at the close of the nineteenth century attracted tourists to Vesuvius in
increasing numbers. The volcano seemed to have been tamed at last when
the travel agent, Thomas Cook, opened a funicular railway up to the summit
in 1880, which was linked to the Circumvesuviana system by a rack-and-
pinion railway in 1902. However, any taming of Vesuvius proved temporary.

The eruption of 1906

Some of the most famous scientists in early twentieth-century volcanology
– Mercalli, Lacroix, Johnston-Lavis and Perret – studied the eruption of
Vesuvius in 1906. Their research revealed the intricate details of the behaviour of the volcano to a widening audience. Vesuvius reciprocated and made
them even more famous.

Vesuvius gave ample warning of its revival. From 20 March 1906, the discharge of the thermal springs in Torre Annunziata became weaker. Then the water levels in the local wells gradually fell by 20–30 cm. A more telling warning came from the northern coast of the Bay of Naples. The shore rose by 10 cm at Pozzuoli in the west, by 20 cm at Naples, by 48 cm at Portici, nearest to the volcano, and by 30–40 cm at Torre Annunziata, in the southeast. These measurements showed that the land was bulging up – especially near Vesuvius itself. The implication was, or should have been, that molten rock was rising into the body of Vesuvius. On 2 April, the telephone poles between Torre Annunziata and Naples began to give out "a noise like a cauldron of pitch boiling violently". On 3 April explosions boomed from the central crater. At about 5.30 a.m. on 4 April 1906, a new fissure opened at a height of about 1200 m on the southern side of the main cone and suddenly gave off a small lava flow. Many strong tremors shook the base of the cone, and lavas and then solid fragments were ejected. At about 2 p.m., more violent explosions began to break up the small cone lying within the main crater.

During the night the wind blew dark-grey ash as far as Naples. At about midnight, a lower vent opened at a height of about 800 m, which soon became much larger than the first. Fluid lava gushed from it with such vigour that the snout of the flow soon travelled 2.5 km from the vent. During that night, too, another fissure burst open on the northern flanks of the cone

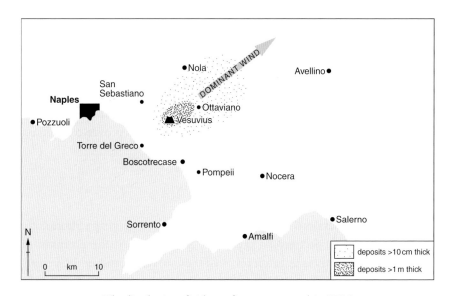

The distribution of airborne fragments erupted in 1906.

almost diametrically opposite the first two vents. On the morning of 5 April, the funicular railway made its last trip up the mountain. On board was the American amateur volcanologist, Frank Perret, who was returning to the observatory. As the morning progressed, vast puffs of ash and fumes exploded from the crater as if they had come from a giant steam locomotive.

At about 8 a.m. on 6 April, a third vent opened on the southwestern flank at a height of about 600 m near Bosco di Cognoli and emitted lava that was unusually fluid and full of steam. Frank Perret and Raffaele Matteucci went down to assess the threat to the settlements below. The lava was attacking the lower slopes on three fronts. The flow had divided into three streams: one started off towards Torre Annunziata, destroying woodlands and crops on the way; another branch travelled towards Boscotrecase; and a third made off towards Terzigno, where it caused less damage because it crossed flows that had erupted in 1834 and 1850. At about 3 p.m. on 7 April, the lava stream travelling towards Boscotrecase stopped at the cemetery about 500 m from the town. The citizens assumed that the eruption was over and they rushed out to inspect the snout of the flow. Those who had removed their furniture from their homes began to take it back again. However, the eruption was not to reach its climax until the following night.

From about 4.30 p.m. on 7 April loud explosions shook all the windows in Naples. That evening, frequent explosions from two vents spurted out great fountains of glowing spatter, ash and cinders to a height of 2 km, and covered the upper 200–300 m of the cone in a spectacular cape of fire. Then the eruptions changed character and huge clouds of black ash and stones were wrenched from the walls of the vent and thrown into the air along with the glowing materials. The explosions cored out a deep crater at least 500 m in diameter. Throughout 8 April, the whole volcano was enveloped in what Johnston-Lavis called an "impenetrable cloud of cocoa-colour dust" that obscured the detail of events. However, most of the fragments fell to the northeast of the cone, not, it seems, blown by the wind, but because they had been hurled from the crater at an angle. Soon, almost 1 m of ash, sometimes including large boulders reaching 10–20 cm across, had accumulated beyond the crest of Monte Somma.

The thickest falls occurred between 2 a.m. and 3 a.m. on 8 April, and weighed heavily on the rooftops. Ottaviano and San Giuseppe suffered especially badly, and practically all the roofs in the two communities had caved in and killed dozens of people by mid-day on 8 April. About 30 roofs also gave way in Somma–Vesuviana. But the greatest single disaster of the whole

Frank Perret describes a close-up of the eruption in April 1906

Toward 10 p.m. [on 7 April] masses of cone-material again began to be mixed with the incandescent jets, with consequent renewal of powerful, brilliant electrical discharges in wondrous contrast to the golden spear-heads of the ejected lava piercing the dark detritus-clouds [ash clouds] . . . [At 10.20 p.m. and 10.40 p.m.] two brilliant fountains of fire upon the flank of the mountain proved to be a violent renewal of activity at the lava vents of Bosco di Cognoli and the Valle dell'Inferno. From these there issued veritable torrents of lava, which in a few hours reached the first houses of Boscotrecase and later crossed the town and the Circum-Vesuvian railway . . . At 12.37 a.m. on 8 April, a strong shock was felt, followed soon after by visibly greater explosive activity at the crater . . . There was . . . a continuous earthquake, and for some hours . . . it was impossible to stand quite still . . . the mountain was pulsing and vibrating . . . The obvious peril to the observatory drove us out-of-doors, where at only 2.5 km from the radiant column of incandescent pasty fragments – at this time already 300 m in diameter and 3000 m in height – the air was so cold as to constrain us to build a fire and sit close about it for warmth. This condition was caused by the aspiration of cold air from the sea by the tremendous updraft of the pillar of fire . . . At 3.30 a.m. on 8 April, there began the truly dynamic culmination of the great eruption, with a literal unfolding outwardly of the upper portions of the cone in all directions, like the petals of a flower. Those who speak of the mountain having fallen in were not eye-witnesses of what occurred. No mass of matter, however great, could descend against the mighty uprush of gas that was now liberated from the depths . . . This colossal column, with ever-increasing acceleration, was actually coring out and constantly widening the bore of the volcanic chimney . . . The fragments . . . now began to fall from the skies till it became necessary always to stand erect and with a rolled-up overcoat held as a cushion upon one's head. At first the size of small nuts, the projectiles gradually increased up to 2–3 kg, some of which were thrown . . . 4 km from the crater.

eruption occurred at about 6 p.m. on Palm Sunday, 8 April, when the roof of the Oratorio church in San Giuseppe collapsed and killed some 105 of the people sheltering within it. Johnston-Lavis, a strong critic of the Campanians, commented that many people had been killed because they had "collected in the churches to pray under the weak and ill constructed roofs, instead of getting on to them to clean them [because], even during the heaviest part of the fall, persons could move about if they covered their heads and shoulders with pillows, tables, and other such improvised shields."

Eventually, the ash and fumes rose to a height of 13 km before the column waned during the evening. To Giuseppe Mercalli, watching from Torre Annunziata, the column was composed of "enormous globular masses like gigantic cauliflowers" in ever-expanding billows of black, grey and white. At the same time, deluges of muddy rain dowsed Naples and Torre del Greco between 8 and 11 April. The water made the ash abnormally heavy and thus undoubtedly contributed to the collapse of the roofs and the consequent loss of life in Ottaviano and San Giuseppe. On the western slopes of Vesuvius, the ash and cinders destroyed the funiculare and seemed about to attack the nearby observatory, but the director, Raffaele Matteucci, and his voluntary assistant Frank Perret, decided to stand firm.

Nevertheless, the chief threat on 8 April was another lava flow that had been advancing towards Boscotrecase since 10.45 p.m. on the previous evening. In a few hours, this flow reached the town, destroyed the Oratorio quarter, invaded the church of Santa Anna, trapped and killed three old men, crossed the road to Torre Annunziata, filled a 500 m-long cutting of the

Sketch, based on a photograph by Lacroix (1906), of the snout of the lava flow invading a street in Boscotrecase in 1906. The flow progresses only slowly as it solidifies, especially in a built-up area, and its snout typically becomes steep and clothed in cinders.

Circumvesuviana railway to the brim, and turned towards Torre Annunziata itself. Then, at 8 p.m., the lava stopped erupting from the vent as suddenly as it had started, and the snout of the flow soon came to a halt at the cemetery in Torre Annunziata.

Cocoa-colour ash fell on 8, 9 and 10 April, followed by greyish ash on 11 and 12 April. On 12 April, the local newspaper, *Il Giorno*, unjustly criticized Raffaele Matteucci for not being as diligent as Luigi Palmieri, his predecessor at the observatory. The newspaper also revealed a communication problem that was to become more acute as the century progressed: common mortals found Matteucci's communiqués incomprehensible and frightening. True to journalistic form, however, on the very same day *Il Giorno* was able to publish a more comprehensible and reassuring interview with "the illustrious volcanologist" Professor Giuseppe Mercalli.

Then on 13 and 14 April, as if to prove Mercalli's reassurances, Vesuvius mantled the whole volcano with the white ash that always promised the end of the eruption. In fact, the eruption rumbled on spasmodically until about 23 April. By then it had become the second longest "final eruption" after that in 1794.

Unfortunately, as in 1822, the end of the eruption did not mark the end of the destruction. On 27 and 28 April, heavy rain set in motion the thick mantle of unconsolidated ash covering the slopes of the volcano, notably – as usual – on the northern slopes of Monte Somma. The resulting mudflows smashed two bridges on the Circumvesuviana railway, covered crops, vineyards, orchards, woodlands and pastures, and swamped Pollena, Cercola and Terzigno. They stripped away the upper two thirds of the funicular railway, and the restaurant, station, machine house, boiler and stables vanished without trace. The cable rolled up like a tangled ball of string, in a jumble of rubble, twisted rails and wires. So much debris accumulated at the foot of the cone that the shoulder, which had marked the southern rim of the Somma caldera for centuries, could scarcely be distinguished at all.

As late as 18 May, another mudflow damaged Resina, and on 21 May yet another swamped San Sebastiano and Pollena. These *lave d'acqua*, as the local people called them, were often as much as 1 m deep. Whenever heavy rains affected the district, the prospect of mudflows thus remained a major threat long after the eruption itself had ceased (and Portici and Torre del Greco were among the towns afflicted even as late as October 1908). As a result, the mudflows provoked some of the first permanent mitigating measures undertaken on the volcano. Uncemented walls of cinders and lava blocks

His Majesty "visibly moved"

Charitable organizations and the Italian government took steps to bring aid to the stricken area, possibly stimulated by the increasing criticisms that had been levelled at the central government from Campania ever since the establishment of the Kingdom of Italy in 1861. The situation was deemed to require a royal presence. Thus it was that, as dawn broke on 12 April, King Vittorio Emanuele III, who in 1900 had succeeded his father, Umberto I, left the royal palace in Naples by motor-car to visit the stricken areas. He was followed an hour later by Queen Elena. As *Roma* reported, by the time they reached the outskirts of Somma–Vesuviana the ash was so thick that the royal motorcar could make no further progress. The king left the vehicle and set about inspecting the damage to the town. Some people recognized him and threw themselves, weeping, at his feet and begged for help for their poor desolate land. The king promised that help would be provided with the utmost urgency. Vittorio Emanuele then took a horse and rode off to Ottaviano under the hail of ash. At 9.30 a.m. he reached Ottaviano, where he was "visibly moved" by the dreadful destruction in the town. As *Roma* put it, "Not a roof had survived; and from every pile of ruins came the cries of the dying; and from every corner came the shouts of the survivors looking in the debris for a lost relative. Near the station, dying victims who had been pulled from the ruins were lodged in the railway coaches".

Still "visibly moved by the piteous sight", the king toured the town, followed by his retinue and a throng of citizens. And again "many survivors implored the king through their tears to bring them assistance. His Majesty had a kind word of encouragement for each one of them, saying that the aid work was proceeding well, in spite of all the difficulties, and that the utmost would be done in the future". Vittorio Emanuele praised the soldiers and the local authorities, but condemned the local priest, who had fled as soon as the catastrophe struck. Still visibly moved, the king rode onwards to San Giuseppe and San Gennariello, where the same pitiable scenes and enthusiastic ovations were repeated. Vittorio Emanuele III then returned on horseback to Somma, where he climbed into his motorcar again and was back in the royal palace in Naples just before noon.

were built athwart any valleys that mudflows seemed likely to follow in future. The authorities then painted them white, no doubt to emphasize that they had done something to protect the population.

The eruption of 1906 caused more damage to the cone than any since 1822. The main cone of Vesuvius had lost 115 m from its summit. In March 1906, the old crater had been full of lava. In May 1906, the new crater was 600 m deep and 700 m across, and its rim rose only to 1100 m above sea level in the east and to 1200 m in the west. Some 218 people had been killed and over 100 had been injured; falling roofs had killed 213, molten lava had killed three people, and emissions of carbon dioxide had gassed another two people a few days after the eruption had all but ceased. The coarse falling ash had badly damaged the vines and fruit trees. Nevertheless, in early May, the vines were already sprouting, and farmers had successfully planted lettuce and cabbages as soon as the heavy rains at the end of April had leached out any noxious salts from the blanket of erupted fragments. Deep ash masked the rough lavas that had been a feature of most of the eruptions of the preceding century. Not only were the lavas suddenly easy to cross, but they carried the promise of a flourishing mantle of vegetation within a decade or two.

The eruption of 1944

After the final eruption of 1906, Vesuvius rested for seven years before resuming its persistent activity on 5 July 1913. Repeated effusions of lava filled the great crater formed in 1906, so that, from Naples, a small cinder cone within it could eventually be seen rising above its rim. This persistent activity was punctuated by five stronger intermediate eruptions, notably during 3–8 June 1929.

Meanwhile, Mussolini had taken charge of Italy in 1922, and in 1940 had led the country into war on the side of the Axis powers led by Nazi Germany. The Allies invaded Sicily on 10 July 1943 and a new Italian government signed an armistice in September. The Allied forces then advanced as liberators through Campania in October 1943. In March 1944, Vesuvius added to the chaos and misery in liberated Naples by embarking upon a "final eruption".

Giuseppe Imbò, the director of the observatory, had stayed at his post for most of the war. When the Allied armies reached Campania, they set up an airbase at Terzigno and took over the tower of the observatory as a radio transmitter. Nevertheless, with the customary devotion of its directors, Giuseppe Imbò stayed at the observatory, even when the building was shaking "like a boat in a rough sea" during the climax of the eruption.

Vesuvius began to show signs of renewed vigour on 13 March 1944, when the small cone in the centre of the crater collapsed. Imbò walked down to tell the military commander of the Allies at Terzigno that the B-25 planes at his base might soon be in danger from more than German bombs. Imbò's advice fell on deaf ears, but his predictions were fulfilled within a week when volcanic ash covered the planes.

The eruption began at 4.30 p.m. on Saturday 18 March, when masses of

The distribution of airborne fragments erupted in 1944.

lava emerged from the crater and formed two flows, one directed southwards and the other northwards into the Atrio del Cavallo. The northward flow was larger, lasted longer, and was the most destructive of the pair by far. On the evening of 19 March, the northern flow surged out from the Atrio del Cavallo towards Cercola, Massa and San Sebastiano. Somehow, in spite of the war and the chaos of the occupation, the Allied authorities commandeered ten coaches to evacuate the 12000 people from these localities. But the lavas quickly developed the typical carapace of solidified clinker that slowed them down to walking pace. The lives of the inhabitants in the three towns were therefore no longer threatened, although their homes and possessions remained in grave danger. Hence, many townspeople returned to their homes to salvage their belongings, and more importantly, to protect them from looters.

As the lava flows were advancing, a huge column of ash and fumes soared into the air above the volcano. The writer Norman Lewis, who was then a British military intelligence officer, recorded his impressions in his diary:

Today Vesuvius erupted. It was the most majestic and terrible sight I have ever seen or ever expect to see. The smoke from the crater slowly built up into a great bulging shape having all the appearance of solidity. It swelled and expanded so slowly that there was no sign of movement in the cloud, which, by evening, must have risen 30 or 40 thousand feet into the sky, and measured many miles across . . . it was absolutely

269

After the eruption in 1906 (600 m deep and 700 m across)

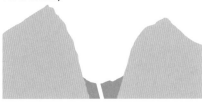

March 1915, after moderate Strombolian activity had started to fill the crater

August 1920, after a small cone had developed within it

February 1944, where the small cone rose higher than the crater rim

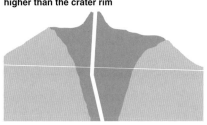

After the eruption in March 1944, 300 m deep and 480 m wide, and situated 200 m to the south of its predecessor

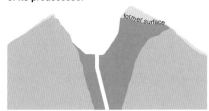

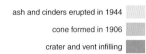

ash and cinders erupted in 1944

cone formed in 1906

crater and vent infilling

Sections showing the changes in the form of the crater of Vesuvius, 1906–1944 (based on illustrations in Imbò 1966).

motionless, not quite painted – because it was three-dimensional – but moulded on the sky; an utterly still, and utterly menacing shape . . . full of swellings and protuberances, like some colossal, pulsating brain . . . [it] trailed uncharacteristically a little tropical liana of heavy ash, which fell earthwards here and there from its branches in imperceptible motion.

Before long, the eruption worsened. A smooth grey pall of ash covered all of Naples, and the military authorities were worried for the safety of their planes and weapons, and probably for their lives, too, although it would have been bad form to admit it. Norman Lewis was instructed to find out what Vesuvius was likely to do next. On 20 March, armed with a suitable gift, Lewis went for "expert advice" from Professor Saraceno, a leading local seismologist, who foretold that there could well be a repetition of the disaster that had destroyed Pompeii. Lewis's gift was more than the forecast was

270

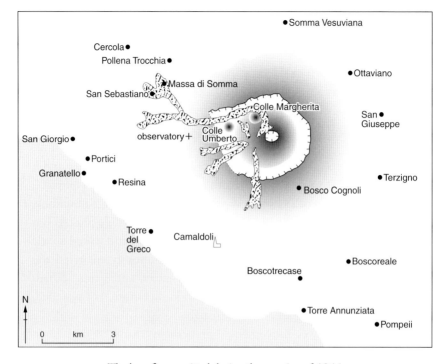

The lava flows emitted during the eruption of 1944.

worth: "I repaid his advice with a tin of corned beef, which he accepted with gratitude".

This gloomy forecast and the growing violence of the eruption increased the anxiety of both the Neapolitans and their liberators. The news then reached Naples that lava was advancing upon San Sebastiano and Massa. San Sebastiano lay in a particularly vulnerable position: in a valley between two flows that had erupted in 1855 and 1872 respectively. To an outsider, its destruction looked inevitable, although, like all those living under the threat of Vesuvius, none of its citizens would even contemplate the possibility. Nevertheless, the flow crept through the vineyards, poking its way into out-houses and setting timber and logs alight, and entered San Sebastiano and Massa in the early hours of 21 March.

On 22 March, Norman Lewis was sent to get an on-the-spot report. Slippery ash and military roadblocks made his journey difficult. Once he arrived in San Sebastiano, both the behaviour of the lava and the attitude of the population towards the crisis made it hard for an outsider to understand:

271

The lava attacked the town in two ways. The more liquid lava poured into and around the buildings and moulded them in its embrace. The more viscous, cooler lava piled up against each obstacle, and crushed and pushed it down like a giant bulldozer. In either case, the progress of the flow was slow, unstoppable and implacable. It seemed to be digesting its prey like some enormous python. I had been prepared for rivers of fire, but there was no fire and no burning anywhere – only the slow deliberate suffocation of the town under millions of tons of clinkers. The lava was moving at a rate of only a few yards an hour, and it had covered half the town to a depth of perhaps 30 feet. A complete, undamaged cupola of a church, severed from the submerged building, jogged slowly towards us on its bed of cinders. The whole process was strangely quiet. The black slagheap shook, trembled and jerked a little and cinders rattled down its slope. A house, cautiously encircled and then overwhelmed, disappeared from sight intact, and a faint, distant grinding sound followed as the lava began its digestion. As I watched, a tall building housing what was clearly the town's smart café took the pressure of the lava's movement. For perhaps 15 or 20 minutes it resisted, then the juddering, trembling spasm of the lava seemed to pass into its fabric, and it, too, began to tremble, before its walls bulged and went down.

Most of the time, under the fascinated gaze of the townspeople, the buildings could offer no more than passive resistance to the lava. Every now and again, however, there would be an explosion when the red-hot lava poured down into a well or oozed over a pool of water and fired jets of steam and droplets of lava into the air. Casks of wine suffered the same, but more saddening, fate. These explosions, hot and unexpected, were the most dangerous and potentially lethal aspects of the flows. That they killed no-one was something of a miracle.

An outsider could not comprehend the attitude of the people, most of whom watched the loss of their homes, goods and chattels with a fatalistic detachment. There was no panic, no wailing, not even many tears. A few citizens gave vent to cries of anguish or anger, and went so far as to threaten the aggressor with retribution. Those who know how volcanoes behave also know that such threats are pointless, and the people of San Sebastiano knew better than most. They could only watch, pray and wait for the volcano to relent. As Norman Lewis recorded:

About 50 yards from the edge of this great slowly shifting slagheap, a crowd of several hundred people, mostly in black, knelt in prayer. Holy banners and church images were held aloft, and acolytes swung censers and sprinkled holy water in the direction of the cinders. Occasionally a grief-crazed citizen would grab one of the banners and dash towards the wall of lava, shaking it angrily as if to warn off the malignant spirits of the eruption. The cinema . . . was still there, protected now by a dozen young men who had formed a line and had advanced, brandishing crosses, to within a few yards of the lava. Not a single clinker tumbled down the black slope as we watched.

Perhaps only the image of their patron saint, San Sebastiano, could help the town now, and his statue dominated all those confronting the advancing flow. But, as Norman Lewis was to discover, he may have had a secret assistant close at hand:

Wandering away down a side street, I noticed another image, also with numerous attendants, which was covered with a white sheet. One of the carabinieri, patrolling on the lookout for looters, told me that this was an image of San Gennaro, smuggled in from Naples on the outside chance that it might be of some use if all else failed. It had been covered with a sheet to avoid offence to the confraternity of San Sebastiano and the saint himself, who might have been expected to resent this intrusion into his territory. As a last resort only, San Gennaro would be brought into the open and implored to perform a miracle. The carabiniere did not think this would be necessary, because it was clear to him that the lava stream was slowing down. Flakes of ash, softer than snow, were still drifting down . . . Childish voices somewhere in the rear had begun to sing a Te Deum. It seemed likely that half the town would be saved.

Meanwhile, as the lava flows had been making their way towards San Sebastiano, the crater had been showering clouds of ash and fumes over the southeastern flanks of the volcano. Then the eruption entered a more spectacular pyrotechnic phase. At 5.15 p.m. on 21 March, a great fountain of clots and droplets of lava gushed for almost an hour from the crater, and often spurted as much as 4 km into the sky. Complete calm then ensued for nearly two hours. At 8 p.m. a second fountain spurted up, and half a dozen similar lava fountains followed at intervals until mid-day on 22 March.

The lava flow erupted in 1944, still relatively free of thick vegetation, tracks westwards along the base of the ridge of Monte Somma.

Thereafter, the character of the eruption changed once again. Great billowing clouds of ash, cinders, fumes and steam rose at least 5 km into the air. Thick showers of ash and small pyroclastic flows rained down onto the southeastern slopes of the volcano. This phase reached its climax in the afternoon of 22 March, but the volcano took a rest between 6 p.m. and 9 p.m., before resuming its violent behaviour until 2 p.m. on 23 March. For the next three days, until 26 March, Vesuvius added more earthquakes to this repertoire of powerful expulsions. But, in fact, its activity was already visibly waning and it gradually came to a halt on 29 or 30 March.

Many homes and public buildings in San Sebastiano and Massa were destroyed, but no victims from these towns were apparently reported from the onslaught of the lava. The deaths came from the falls of ash on the southeastern flanks of the volcano. Ash had accumulated thickly here and 80 cm were recorded from Terzigno and Pompeii, for example. Not for the first time, the accumulations of ash on the rooftops provided the greatest threats and caused the fatalities. Collapsing roofs killed 12 people in Nocera and 9 people in Pagani. (The Allied armies reported another 21 deaths from unspecified locations, which may have included the deaths noted by the Italian authorities, which would thus be counted twice.) Large clumps of falling clinker killed three unfortunates in Terzigno, and noxious gases also took their toll when two people were asphyxiated, probably by carbon dioxide,

***Vesuvius in 1944 (courtesy of Derek Hopper and Tom Scotland,
then of 614 Pathfinder Squadron, Royal Air Force)***
*(a) Lava flows invading San Sebastiano. (b) A lava flow toppling, and setting fire to, a tele-
graph pole near San Sebastiano.* **(Continued overleaf)**

Vesuvius in 1944 (continued)

(c) View of Vesuvius from the north. (d) A close-up of the crater of Vesuvius in a moment of relative calm during the eruption.

after they took refuge in an air-raid shelter in Resina on 24 March. The death toll was certainly 26, but it may have reached 47.

One major consolation of the eruption in 1944 was that the great accumulations of ash occurred to the southeast of the volcano, whereas the lava flows occurred to the northwest. Had the two forms of the eruption happened in the same area, then more lives would probably have been lost and the evacuation would have been severely limited.

The eruptions of 1872, 1906 and 1944 each expelled roughly the same volume of materials. In 1944 they amounted to some 70 million cubic metres of lava flows and fragments. The eruption in 1944 also removed most of the material that had gradually choked the crater since 1913. At the close of the eruption, the summit of the cone was distinctly asymmetrical: it rose to 1281 m in the north, but reached only 1169 m in the west. The new crater had also shifted 200 m to the south, but it was only one third of the size of its predecessor: 300 m deep, 580 m long and 480 m wide. Subsequently, the crater has often been the scene of variable mild fumarole activity, but there has been little sign of any revival of lava eruptions.

As Norman Lewis commented, the effects of the eruption on Naples added to the psychological traumas brought about by the chaos, lawlessness, unemployment and food shortages after the liberation of the city. Sometimes the disorder expressed itself in undisguised criminality, but at other times in greater religious fervour than was common even in Naples. Every Neapolitan yearned for the consolation of cures, marvels and miracles; and they were duly reported from all over Campania and farther afield.

Further reading

Imbò 1949, 1965; Johnston-Lavis 1909; Lewis 1978; Mastrolorenzo et al. 1993, 2002; Palmieri 1872, 1873, 1887; Palumbo 2003; Perret 1924.

Chapter 12

The Campi Flegrei: an eruption that failed

Two episodes of rising ground and earthquakes around Pozzuoli aroused fears of an eruption that did not, in fact, occur. However, the studies undertaken mean that the authorities should be better prepared next time.

Just as attention turned to the volcanic aspects of the Campi Flegrei while Vesuvius lay dormant between 1139 and 1631, so the Campi Flegrei have again been the focus of renewed scrutiny since Vesuvius fell silent in 1944. Its activity occurs on a more modest scale and has none of the spectacular bombastic quality of such a powerful volcano as Vesuvius. Nevertheless, the volcanic features have had an impact on human society that far exceeds their actual accomplishments. Sulphurous emissions had given La Solfatara its name ever since antiquity, and Earth scientists adopted the term "solfatara" as the generic name for such hydrothermal activity throughout the world. More importantly, however, during the later decades of the twentieth century, bradyseismic movements have twice threatened – but have so far failed – to cause an eruption in the densely populated area in and around Pozzuoli.

La Solfatara

La Solfatara has been the most consistent performer among all the volcanoes in the Campi Flegrei for the past 2000 years, and it has played major roles in folklore and legend. More than any other volcano in the Campi Flegrei, La Solfatara contributed to the reputation of the region in antiquity as the "burning lands" and as a major gateway to the Underworld. Later, among the Christians, it seemed to offer a small sample of the hellfire and brimstone with which the Church always threatened sinners. Of course, La Solfatara

The Piano Sterile, the floor of the crater of La Solfatara, a hot wasteland of fumaroles and occasional mudpots.

had always been accessible, easily visible and very well placed among a large and literate population, but its fame is no mean achievement for a volcano that has expelled no lava flows, no cinders, little or no ash or pumice, and has produced only the very occasional brief geyser since the eruption that gave birth to it about 3800 years ago. Its contributions to volcanic activity have been mild: whispy diaphanous fumes swirling from sulphurous holes, and water bubbling and hissing in small pits in the grey floor of the crater.

La Solfatara is a horseshoe-shape cone with a deep rectangular crater, the Piano Sterile, which is wide open to the southwest. This almost flat arena, 750 m long and 600 m broad, is the centre of its hydrothermal activity. This activity develops because the shallow magma heats the rocks beneath the volcano, which in turn heat any rain that sinks into the ground, so that the water returns to the floor of the crater in bubbling pools, mudpots and hissing fumaroles and solfataras exhaling steam and sulphurous fumes. The sizes and positions of these pools and conduits change frequently in response to alterations or blockages in the subterranean plumbing system, which are brought about primarily by earthquakes. At present, several vents cluster on the eastern half of the Piano Sterile and the crater walls near by. Each cluster has its individual characteristics: the Forum Vulcani, for instance, is small and vigorous, whereas the Bocca Grande is the largest and hottest of all. Nearest the centre of the crater, the Fangaia area has pits of hot mud that are commonly 2 m deep and up to 500 m^2 in extent. As well as steam, boiling water and hot mud, these vents also give off carbon dioxide and hydrogen sulphide, and many are encrusted with beautiful sulphur crystals that grow when hydrogen sulphide oxidizes upon reaching the air. The sulphur can then react with water to form sulphuric acid, which bleaches and chemically alters the rocks to form the bianchetto, the grey mantle of hard mud that floors the crater. There is a pervading smell of sulphurous compounds.

If the vents operate for a long time, they build up mounds of hardened mud that rise 2–3 m between the new muddy pools that appear around them from time to time. The less active areas are coated with greyish-white hardened mud that resonates underfoot. However, sometimes this crust is dangerously thin, and visitors are well advised to follow the safe catwalks that have been prepared, in order to avoid falling into the boiling waters just below the surface. Overall, however, these threats are relatively minor. Nevertheless, the dangers in the crater will be of an entirely different order if ground movements in recent decades are anything to go by. These movements suggest that the crater of La Solfatara, or a spot near by, could well become the source of the next eruption of magma that will form a new volcano in the Campi Flegrei.

Fumes issuing from the foot of the northern wall of the crater of La Solfatara.

Bradyseismic movements

Ever since the Neapolitan Yellow Tuff exploded and formed its caldera some 12 000 years ago, volcanic activity within the Campi Flegrei has tended to become concentrated towards the centre of its caldera. The eruptions have also become less powerful.

The rising bradyseismic movements of the ground are potentially more threatening, because they carry the menace of a volcanic eruption, like the one that formed Monte Nuovo in 1538 after the ground had been rising for at least four decades. Then, as soon as the eruption ceased, the land began to sink back down once again. This subsidence continued into the twentieth century, and Pozzuoli seemed to lie at its hub. Measurements of the Roman market (the Temple of Serapis) at Pozzuoli from 1820 until 1969 showed that it had sunk by a total of 2.1 m. The average rate of 1.4 cm a year suggested that the total subsidence at the site since 1538 could have amounted to more than 5 m. This subsidence implied that the magma had drained back, perhaps several kilometres below the surface, and that another eruption was unlikely to occur as long as this trend continued.

Late in 1969, a railway worker noticed that a section of the Flegrean Railway had been raised up, and at the same time, citizens of Pozzuoli observed that parts of the town also seemed to have risen. Accurate measurements soon confirmed that another phase of ground uplift had indeed begun. If another eruption was on the way, then 400 000 people in the district were in danger of finding a volcano on their doorstep.

The geologist Franco Barberi and his colleagues began to study develop-

Pozzuoli in 1966 before the latest sets of bradyseismic movements began and where the surface of the quays lies close to the sea level.

ments. It so happened that the shoreline at Pozzuoli had been accurately levelled in 1968, but a new survey in March 1970 showed that the land had risen by an additional 98 cm already. More surveys showed that the land continued to rise until July 1972. The maximum uplift of the ground occurred in Pozzuoli itself, which was 1.70 m above the level recorded in 1968. However, the land then sank back by about 20 cm over the next two years, so that the net maximum rise was 1.50 m by the time the movements stopped in 1974.

Hundreds of earthquakes accompanying these bradyseismic movements were recorded on the local seismographs, but relatively few were strong enough to be felt by the inhabitants. Nevertheless, most people were deeply anxious, when they were not totally panic stricken, that a large earthquake or a volcanic eruption (or both) was about to strike them. Many thousands of townspeople left Pozzuoli. It was just as well that the feared calamity did not befall them, because the authorities had not been keeping the area under systematic earthquake and volcanic surveillance, and they had devised no civil-defence plans to evacuate people if ever such an emergency were to arise. It was strange, if not entirely inexplicable given the local society, that one of the most densely populated volcanic areas in the world should have been protected so inadequately.

Some of these omissions and deficiencies were corrected during the ensuing period of bradyseismic calm. Seismographs continually monitored earthquakes; repeated precise levelling and tide-gauge surveillance monitored ground movements; gravity measurements monitored possible ascents of magma; and chemical analysis monitored changes in the contents of fumaroles and hot springs. All these methods were co-ordinated and analyzed in the research centre at the Vesuvius observatory.

The civil and the scientific authorities were therefore better prepared when the next crisis came in July 1982. Ironically, the signs given out were ambiguous and they put the experts in a quandary. The ground started to rise, especially at Pozzuoli, but at first there were no abnormal earthquakes. The vents at La Solfatara started to bubble more actively but then calmed down again, suggesting perhaps that the magma might not be rising after all. However, the earthquakes finally began in November 1982, and indicated that the magma was indeed rising. Most of the earthquakes were centred along the shore near Pozzuoli at depths usually ranging 3–4 km below the surface. The two strongest earthquakes were of relatively low magnitudes: 3.5 on the Richter scale at La Solfatara on 15 May 1983, and 4.0 at Pozzuoli on 4 October 1983. They caused so much panic in and around Pozzuoli that the

The uplifted quays at Pozzuoli. The cars are parked on the new lower quays, and the old higher quays rise above them and in the foreground.

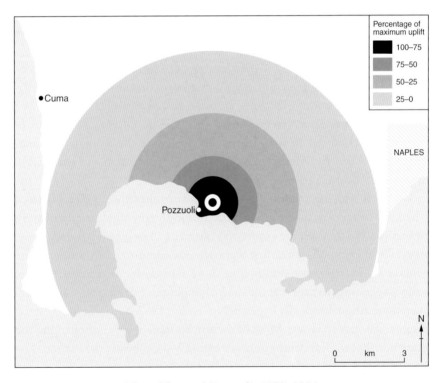

The uplift around Pozzuoli, 1982–1984.

authorities decided to evacuate 40 000 people from the town. In an efficient resettlement programme, a new town capable of housing 57 000 people was constructed in six months at Monte Rusciello. This evacuation probably saved many citizens from injury, because swarms of tremors caused the weaker buildings in Pozzuoli to crumble and sometimes to collapse, especially between October 1983 and April 1984. An eruption seemed inevitable. But, in fact, the foci of the earthquakes showed no sign of rising to the surface, which implied that the magma apparently associated with them was not rising either. No rising magma: no eruption – unless the situation underground were to change.

The whole region remained on tenterhooks until December 1984, when the ground stopped quaking and the land stopped rising. The feared eruption did not take place. But Pozzuoli and its citizens did not escape unscathed. As in the previous crisis, Pozzuoli had suffered the greatest uplift, and no less than 8000 buildings in the town had been damaged. The harbour area had risen by 1.85 m. Thus, the centre of Pozzuoli had undergone a net rise of 3.35 m after the summer of 1969. One major consequence was that the harbour became too shallow for large boats, and another was that the quays alongside the harbour had been lifted out of the reach of small craft, so that a lower landing stage had to be built to cater for them.

The ambiguous and inconsistent signals given out during the two bradyseismic crises have challenged Earth scientists, who have been led to question even the basic causes of the movements. Some have suggested that the uplifts of the land were caused by pressure changes in a body of water in an aquifer subject to deep-seated temperature changes. However, most research workers have been more pessimistic. They believe that the movements are indeed related to displacements of magma, and that the two most recent bradyseismic crises will have developed many new fissures in the rocks. Hence, an eruption may well be inevitable next time, because the next batch of rising magma will find a much easier path to the surface. Moreover, enough raw material is still available for another eruption, because a small body of magma, about 1.4 km^3 in volume, lies only 3–5 km below the surface.

Planning for the next eruption

Some experts have suggested that an eruption could occur in the Campi Flegrei before Vesuvius resumes hostilities, but that such an eruption would probably be weaker than the one that formed Monte Nuovo. This would normally be good news for the inhabitants, except that the next outburst may well occur in the very centre of Pozzuoli or less than 2 km away near La Solfatara, which lies virtually in its suburbs.

The history of the Neapolitan Yellow Tuff caldera suggests that few, if any, lava flows would be emitted, and would present little mortal danger if they were. The most dangerous element of the next eruption would be small pyroclastic flows, similar to those of 1538, which could surge as much as 2–4 km from the vent. However, it is impossible to guess where they might emerge, or which direction they will travel, although it seems almost inevitable that the most severely affected zone would be around Pozzuoli. Ash and pumice would probably fall up to about 3–4 km from the vent, but the wind could carry the finer materials more than 10 km away. Their distribution would be determined by the dominant wind at the time, although it generally blows from the western quarter in the Campi Flegrei. Because the experts cannot anticipate the directions that any of these threats will take, any pre-emptive evacuations have to be planned on the assumption that *any* sector around the anticipated vent could be affected, which of course greatly multiplies the numbers of possible evacuees.

It is anticipated that a core area within a radius of 2–4 km from the vent could be devastated. This area houses some 350 000 people, with a density of almost 5000 per km^2. Any number of them could be killed if they were not evacuated before the eruption started, and it would probably take three

Modern Pozzuoli from the summit of Monte Nuovo. For centuries, the old town of Pozzuoli was restricted to the small headland in the middle distance, but recent urban and industrial developments have spread both northwards and westwards along the shore – into the very area most likely to be damaged during the next eruption in the Campi Flegrei.

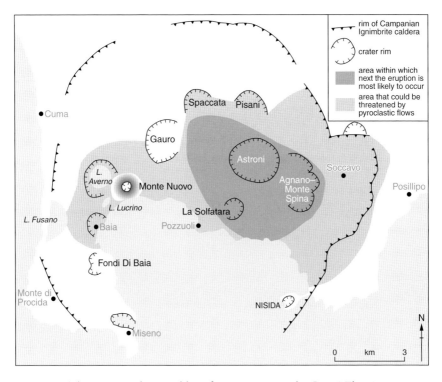

The area most threatened by a future eruption in the Campi Flegrei.

or four days to clear the area. A zone surrounding this core could also suffer from lethal pyroclastic flows, and dangerous ash and pumice falls. Small tsunamis occurred when Monte Nuovo erupted in 1538, and there is an outside chance that they could cause damage along the shore.

Bradyseismic earthquakes would present a major, and perhaps more widespread, threat in the Campi Flegrei, where they have already damaged buildings around Pozzuoli – even without an eruption – during the two most recent crises. Pozzuoli would be especially vulnerable again, because so many buildings in the town are old and do not have anti-seismic protection. The earthquake damage could extend as far as Miseno in the west and Naples in the east.

Further bradyseismic movements in the Campi Flegrei are inevitable, but, as they have shown little regularity, it is hard to extrapolate the pulsations in order to forecast when they will resume. It is anyone's guess when, or even if, they will cause an eruption.

**Planned alert levels for the Campi Flegrei
(Dipartimento della Protezione Civile)**

Level of alert	State of volcano	Probability of eruption	Likely time of eruption	Action and communications
Base level alert	No significant variation of monitored indicators	Very low	Indefinite: not less than months away	Surveillance according to normal Programme
Caution	Significant variation in monitored indicators	Low	Indefinite: not less than months away	Surveillance increased: Vesuvius observatory issues daily bulletins
Pre-alarm	More variations in monitored indicators	Medium	Indefinite: Not less than some weeks away	Continuous surveillance and simulation of possible features of eruption. observatory continually communicates state of volcano to the Dept of Civil Protection
Alarm	Appearance of eruptive features and their monitored indicators show that an eruption is likely	High	From days to months away	Surveillance by remote control. Observatory continually communicates state of volcano to the Dept of Civil Protection

Municipalities in danger from an eruption in the Campi Flegrei and the regions that might receive the refugees in the event of a volcanic eruption

Municipalities (with their populations)	Proposed twinned receiving regions
Naples: Pianura (60 000)	Emilia Romagna
Naples: Vomero; Arenella; Chiaia (27 000)	Apulia
Naples: Bagnoli (27 000)	Basilicata
Naples: Soccavo (51 000)	Calabria
Naples: Fuorigrotta I (40 000)	Lazio
Naples: Fuorigrotta II (40 000)	Toscana
Bacoli (27 000)	Marche
Pozzuoli (70 000)	Abruzzo
Monte di Procida (14 000)	Molise

Further reading

Barberi et al. 1984; Barberi & Carapezza 1996; Dipartimento della Protezione Civile 1984 *et seq.*; Di Vito et al. 1999; Lirer et al. 2001; Luongo & Scandone 1991; Orsi et al. 1996.

Chapter 13

The future: the eruption to be avoided

The next eruption of Vesuvius is inevitable, but scientific and social contingency plans are in place to mitigate its impact on the population. In a region beset with a multitude of social problems, many local people continue to underestimate the dangers of an eruption.

Vesuvius is a violent volcano lying in a densely populated region. This combination makes it the most dangerous volcano in Europe. The next eruption could cost many thousands of lives unless scientific experts and the local and national authorities can develop foolproof plans to address the emergency. In recent decades, contingency plans have been developed with increasing sophistication, but Vesuvius has not really been helping the preparations to confront its next eruption. From 1631 to 1944, eruptions were so frequent that its short dormant interludes were seldom without an attractive plume of fumes, which declared that Vesuvius could unleash an outburst at any moment. Since 1944, there have been few such plumes and Vesuvius has presided benevolently over the Bay of Naples. Only the elderly can remember the events of 1944, and trees, shrubs, and even buildings, are springing up on the most recent lava flows. Paradoxically, even the bradyseismic crises of 1969–1972 and 1983–1984, which caused panic in Pozzuoli, helped tranquillize the population living around Vesuvius. They expected Vesuvius to react, but it took no notice. (There was, of course, no geological reason why it should have reacted.) The apparent indifference of the volcano merely reinforced the widely held conviction in the region that Vesuvius is extinct. Consequently, those who hold this view feel no need whatsoever to take any part in plans to counter a future eruption. Memories are indeed short.

The past is the key to the future

One of the foundations of the Earth sciences is that the study of the processes acting at present offers the best way to understanding what happened in the past. The best way of forecasting what might happen in the future is also to find out what happened in the past. The enormous bibliography on Vesuvius provides Earth scientists with enough information to establish its general pattern of behaviour, although interpreting this pattern in detail still runs into a whole range of problems. All volcanoes behave with a certain consistency, but their eruptions do not obey rigid timetables. Moreover, those who would forecast eruptions have first to reconcile two very different scales of operation. Volcanoes act on the geological timescale, in which centuries have passed like mere flashes in the 4.6 billion years of the Earth's existence. Human beings live on the historical timescale, in which a century is usually more than a lifetime, and where the variation of an hour in the timing of an eruption can determine life or death for thousands of people. Throughout the world in recent decades, the increasing study of signs warning that eruptions might be on the way, and the development of contingency plans to mitigate their effects, mark some of the first steps that mankind has taken in the unequal contest with volcanoes. The students of Vesuvius have been in the forefront of this combat.

Warning signs

Like every other volcano, Vesuvius has always given out signs warning that it was going to erupt. The volcano has often heralded impending disaster by moderate effusive activity that slowly filled its crater, although the preliminary eruptions may have lasted anything from a few weeks to many months. Less obvious signs commonly proved to be much more valuable indicators. Unfortunately, time and again in the past, the threatened people had no means of knowing what the signs meant; they did not see them as warnings, but only as yet another set of natural scourges sent to try them. These warnings could not be appreciated, and therefore heeded, until the basic mechanisms of volcanic activity came to be understood within the past century.

Rising magma lies at the origin of all such warnings. The land surface swells up slightly around the volcano, the ground cracks, noxious gas and steam escape, wells and streams sometimes dry up, and earthquakes shake the land near the volcano more and more often. Sometimes, as the eruption approaches, the ground shudders incessantly. At best, symptoms such as these must have formed the basis of the old myths that great Titans were imprisoned beneath the volcanoes and were struggling to escape. Until modern times, such beliefs were the nearest that observers usually came to linking these earthquakes with eruptions.

The swelling land surface

When magma rises into the crust under a volcano, it displaces the rocks and makes the land surface bulge upwards. The resulting swelling is very slight and it usually passed entirely unnoticed until very sensitive instruments capable of accurate measurements were installed around the volcano. Occasionally, the swelling could be more evident if it took place around a volcano near the sea, such as when the land bulged up along the northern shores of the Bay of Naples before the eruption in 1906.

Gas, fumes and water supply

The rising magma often opens up new fissures that riddle the rocks. The pressures confining the magma are then reduced as it ascends, and the gases within it separate out and escape to the surface. Such gases knocked out King Ferdinand's dog at Portici in October 1766. In general, when sulphur dioxide, hydrogen sulphide or carbon dioxide begin to emerge, then the magma is on the brink of erupting. The magma also heats up the rocks and the groundwater, which then rises to the surface in hot springs, mudpots and solfataras. If these emissions increase, an eruption is probably more likely.

As the magma approaches the surface, it distends the rocks, and blocks many small conduits of underground water, so that wells and even small streams dry up, as occurred, for instance, around Vesuvius before the eruptions in 1631 and 1822. Such changes become suspicious if they cannot be explained by the weather, and especially if they happen at the same time as other warning signs. Wells were vital water sources, so any fluctuations in water level were usually noticed at once, and by the eighteenth century those around Vesuvius began to regard them as warnings of future eruptions.

Volcanic earthquakes

Magma invades the Earth's crust in intermittent surges, pushing the rocks aside and causing earthquakes. Rumbling and earthquakes have preceded most of the major eruptions of Vesuvius. As the magma rises, so the foci of these earthquakes also rise and become concentrated under the volcano. They are perhaps the most important of all volcanic warning signs, because these foci can now be traced and automatically recorded. As the magma rises, the foci of the earthquakes shown on the seismographs become progressively shallower, and thus enable the experts to forecast, often to within a few days, when an eruption might occur. At length, the magma sometimes causes a kind of volcanic resonance with low-amplitude harmonic vibrations, almost as if it were ringing a bell. This is no time to linger on the volcano: the eruption is about to start.

The experts studying Vesuvius have a fine record of research into volcanic earthquakes. As early as 1795, for instance, Ascanio Filomarino designed a basic seismograph to register the earthquakes around the volcano; and in 1856 Luigi Palmieri built a more sophisticated seismic instrument in the new Vesuvian observatory just after he had become its director. These seismographs revealed tremors far smaller than those that people normally noticed, and therefore gave earlier warnings of impending volcanic activity.

Instrumental surveillance: intensive care

During the past few decades, the world's most dangerous volcanoes have been put under constant surveillance. Instrumental surveillance of a dangerous volcano must be detailed, continuous and sophisticated in order to provide the complete pattern of behaviour that the experts require for constant scrutiny, analysis and interpretation. Since 1970, Earth scientists have enmeshed Vesuvius in a dense network of different instruments that register its every change from every angle, and send out sheaves of information that might reveal when it is about to spring back to life. Such surveillance forms the basis of plans to mitigate the impact of the next eruption, but it offers no guarantee that it will save the whole population.

Nowadays, the seismic information is relayed automatically to the observatory for interpretation. Ground deformation is registered from a dense geodetic network, which can measure any bulging of the land to within 1 mm in 1 km. Geochemical monitoring of gas emissions reveals any increases in discharge and changes in the composition of the gases. These indications are more significant if several warnings occur together and seem to be strengthening, or are happening more often. The experts consider that these signs could first become apparent two or three weeks before Vesuvius erupts, although some pessimists fear that the warning time could be much less.

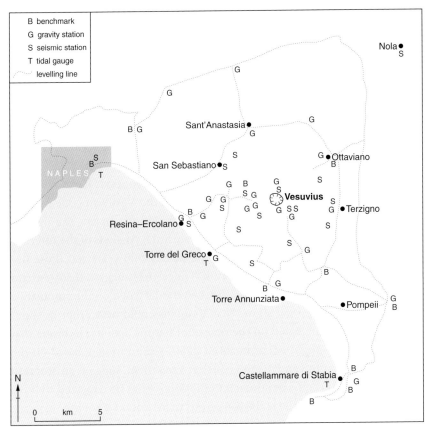

The chief aspects of the surveillance network of Vesuvius.

When will Vesuvius erupt again?

Forecasting when Vesuvius will erupt again is perhaps the most important question in volcanic Europe. A long-term forecast of the date is a funda-mental requirement of all the studies of the future behaviour of the volcano, because the people threatened will have to be evacuated before the eruption starts. Vesuvius will undoubtedly issue plenty of warnings, but it is still im-possible to make an accurate long-term forecast of even the year when the next eruption could occur. Even when the warning signs have started, it will be difficult to predict the exact date, and especially the hour, of a powerful eruption.

In general, the longer Vesuvius rests between eruptions, the more violent is its re-awakening. The possible date and the nature of the next eruption are thus clearly entwined. Vesuvius has now been dormant since 1944, longer than any period since the eruption of 1631 brought 492 years of repose to a catastrophic conclusion. The next eruption, then, is likely to be more violent than any between 1631 and 1944, and could well be as severe as that in 1631 itself. Thus, the longer the present dormant interval lasts, the more violent, the more damaging, and the more widespread the next eruption threatens to be, and the more wide-ranging the precautions will have to be to minimize its impact on the population.

In order to estimate the exact date of the next eruption, Earth scientists have used all manner of statistical tools to crack the Vesuvian code by analyzing the dates and incidence of the previous eruptions. In fact, the answers that they have obtained have varied relatively little. Analysis of the intervals between the Plinian eruptions during the past 25 000 years and those between the more powerful "final eruptions" since 1631 suggests that Vesuvius could next erupt between 2031 and 2064. A model based on the cumulative energy released since 1631 shows that Vesuvius could next erupt in about 2023; another model, based on the revival of seismic activity, indicates that the next eruption might occur in about 2030; a third model, based on the length of the eruptions and the intervals between them, suggests an eruption in about 2040. In each case, the next eruption would be expected to be as violent as that in 1631. Thus, the dangerous years around Vesuvius may start in 2023, or they may be delayed until 2064. Nevertheless, such long-term forecasts still range too widely for them to be incorporated usefully into a contingency plan for the next eruption.

What will be erupted?

Whenever the revival happens, what Vesuvius erupts could be a matter of life and death for the people living on its flanks. The volcanic products that move slowly, such as lava flows, could be diverted and can usually be avoided by refugees; they are major features of moderate effusive–explosive eruptions, and their chief threat is to buildings and crops. Falls of ash and pumice play an intermediate role and occur in both moderate and violent eruptions: they destroy crops, pile up on rooftops and make them collapse, and they present a major hazard to the transport systems that are vital for evacuating the population. Violent eruptions unleash the most lethal elements, such as pyroclastic flows, mudflows and tsunamis, which are all the more dangerous because they move very quickly across the land surface. Pyroclastic flows have been the greatest killers erupted by Vesuvius because they are swift, hard to avoid, and almost impossible to survive. Torrential rains drenching loose volcanic fragments that clothe steep slopes provide ideal conditions for mudflows to develop, and they have caused destruction several times on the northern slopes of Somma–Vesuvius. Tsunamis occurred in AD 79 and 1631. Any contingency plan therefore has to cater for a wide range of volcanic activity, but planners have to prepare for the worst and they are obliged to lay the emphasis on the most lethal and destructive volcanic elements.

The contingency plan for Vesuvius

Although the next eruption ought not to come as a surprise, no-one will be able to stop it, and the scope for palliative action will be severely limited. The crater of Vesuvius lies in a very threatening position, a mere 8 km from the suburban fringes of Naples and even closer to some of the most densely populated areas of Italy. It was therefore vital that a well thought-out proactive plan should be formulated before the next crisis, rather than some off-the-cuff reactive response produced in haste once the eruption has begun. In any case it would then be far too late to act. Any plan would have to involve an evacuation of population on a scale that can scarcely be imagined.

Even with a contingency plan to cope with the next eruption, panic will probably predominate. The terrified people would try to flee to where they believed they would be safe, which, as always, would be the nearest city. Everyone would most probably rush to Naples, creating an enormous traffic jam between the city and the volcano, and effectively preventing any rescuers or aid agencies from entering the threatened area from the west. This is what happened on a much smaller scale in 1631. The impenetrable darkness would exacerbate the general panic. Falling ash and pumice would clog lungs and car engines, and cause slippery roads that would impede the traffic further. Clearly, such total chaos has to be avoided at all costs.

During the second bradyseismic crisis in the Campi Flegrei, the anxieties of the people who had suffered from the earthquakes and the evacuations reached the ears of the government in Rome. In 1991, the Dipartimento della Protezione Civile finally appointed a commission to establish guidelines for evaluating volcanic risks in Campania. The expert members of the National

Volcanology Group (GNV), with the experienced volcanologist Franco Barberi as its president, concluded that Vesuvius would eventually spring back to life with an eruption similar to, and just as powerful as, that in 1631. Therefore, they based their guidelines on that eruption and considered the likely effects of lethal and damaging ash, pumice and pyroclastic flows. Four working groups collected data, evaluated the volcanic risks, developed plans for public-awareness campaigns, and drew up planning strategies for the emergency. In 1995, the GNV outlined the eruptive scenario that they expected, and the commission of the Department of Civil Protection published its first plan for Vesuvius. They believed that, as in 1631, the initial eruptions, accompanied by quite strong earthquakes, would last for several hours and despatch ash and blocks up to 10 km from the summit. The main part of the eruption, lasting at least for several hours, would form a huge eruptive column, perhaps 20 km high, from which ash and pumice would probably spread over some 200–300 km^2 and fragments winnowed from it could make roofs cave in up to 30 km from the vent. During the ensuing hours, the eruptive column would collapse from time to time, sending pyroclastic flows rushing down the volcano with the potential to cover some 50 km^2. During the final phase, lasting weeks or months, mudflows and floods would be likely to develop on the unconsolidated ash and mud on the slopes of the volcano. However, it is perhaps worth mentioning that an outburst as violent as that in AD 79, or the Avellino eruption in about 1780 BC, would cause devastation of a higher order of magnitude.

Hazard zones

As a result of criticisms and the experience gained over the decade, the commission has modified the initial plan and regularly brought it up to date. One of the vital parts of all their deliberations was to identify the vulnerable hazard zones where the people would be in the most danger and where the damage to property and production capacity would be the greatest. The commission established two major hazard zones, and later intercalated a third:

- The Zona Rossa (Red Zone) covers an area of 236 km^2 around the summit of Vesuvius. It includes 18 municipalities and some 600 000 people. It contains the whole volcanic edifice of Somma–Vesuvius, as well as surrounding towns such as Pompeii, Torre Annunziata, Torre del Greco, Ercolano (Resina),[*] Portici, San Sebastiano, Somma–Vesuviana and

---

[*] Resina was renamed Ercolano in 1969.

Ottaviano. Many pyroclastic flows, mudflows and heavy falls of ash, pumice and large blocks could probably destroy nearly everything and everyone within this zone. (Pyroclastic flows, for instance, devastated some 40 per cent of this area in the eruption of about AD 472, and 20 per cent of this area in 1631.) However, no-one can yet anticipate which areas Vesuvius will affect next time. Hence, every municipality within a radius of some 10 km of the crater has to be placed in the Zona Rossa, and the population will have to be removed from it before the eruption starts.

- The Zona Gialla (Yellow Zone) covers an area of 1125 km^2 around the Zona Rossa. It includes 96 municipalities and a population of about 1 100 000 people in the provinces of Naples, Avellino, Benevento, and Salerno. Ten per cent of this area was badly damaged in 1631, and it would be expected to suffer a similar fate during the next eruption. The demarcation of this zone is mainly determined by the likely distribution of wind-blown ash and pumice (which could cause many roofs to collapse), as well as of mudflows and floods. Again, this is not a simple matter, because no-one can forecast the direction of the predominant wind when the eruption occurs. Therefore, a very wide area has to be included in the Zona Gialla. Primarily because it is believed that the winds would take the airborne fragments broadly eastwards, Naples is deemed to lie outside this zone. However, if the wind were to be blowing in the opposite direction when the next eruption occurs, then Naples might face a calamity.
- The Zona Blu (Blue Zone) was added later. It covers an area of about 370 km^2 and includes 14 municipalities and 180 000 people living in and around Nola, north of the Somma ridge. This zone suffered from destructive floods and mudflows in 1631 and would experience the same problems again after each storm until the fragments became consolidated.

Once the threatened areas had been broadly outlined, the next planning task was to decide what could be done when the eruption loomed. The only remedy seemed to be mass evacuation, which, in the Zona Rossa at the very least, would probably have to be undertaken before the eruption started. Thus, in 1993, another commission was set up to prepare such an emergency evacuation. The resulting plan established alert levels for the next eruption. It also outlined instructions about what should be done when each level was reached; about the means of informing and educating the threatened population; and about methods of relaying the scientific conclusions effectively to the people via the local and national administrators, and to the media.

Maps illustrating how all these threatened towns are vulnerable to both

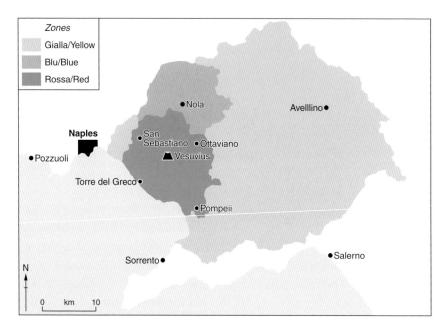

The three danger zones around Vesuvius: Gialla, Blu and Rossa. The boundaries of the zones follow those of administrative units to facilitate plans for an evacuation, and thus give only a general idea of the exact limits of the possible volcanic danger.

earthquakes and eruptions were drawn up; schoolteachers were given courses explaining the problem, so that they would know exactly what to do in the emergency. It was assumed, no doubt, that these teachers would provide the most reliable link with children, their parents and the civil-defence author-ities. It was decided to update the plan at regular intervals.

The experts at the Vesuvius observatory were made responsible for defin-ing and issuing the alert levels as the instrumental data were received. The commission prepared a plan for one threatened town to show an example of how an evacuation should be carried out, with designated assembly and departure points and refugee reception regions, with which each town was twinned. In theory, it could then be used as a prototype for the evacuation of similar settlements around Vesuvius. A preliminary small-scale exercise was then carried out on Sunday 21 November 1999, when 600 out of the 24 000 inhabitants of Somma–Vesuviana were evacuated in 14 buses to Abruzzo, the town's twin refugee reception region. A similar exercise was carried out in Portici in 2001. On 22 October 2006, a larger exercise, called Mesimex 2006, was carried out in the towns in the Zona Rossa, and involved a total of 2000 people.

Mesimex 2006

On 22 October 2006, a major emergency simulation exercise (Mesimex 2006) was undertaken with much publicity and dissemination of information to increase public awareness of the details of the contingency plan. It was supported by three exhibitions aimed at all sections of the community. They made extensive use of pictures, photographs, engravings, comic strips, historical documents and a multitude of scientific investigations to demonstrate how the threats from Vesuvius have been permanent and interrelated during the past 300 years. However, the heart of the exercise was a simulation of an evacuation from the municipalities in the Zona Rossa. A total of 2000 people participated from every municipality except Portici, where the municipal authorities claimed that they had already learned enough during their exercise in 2001 (and thereby revealed another difficulty facing those planning a series of such exercises). A further major feature involved the extent to which the planners themselves, and 15 expert foreign observers, could learn from co-ordinating the exercise in the field. The protection of the rich cultural heritage of Campania was also considered, with a simulation of the effects of fire on a Roman villa at Somma–Vesuvius. The people selected for evacuation formed a cross section of the local population and included nuns, unmarried mothers with their children, and a group of young people playing the role of tourists who had been taken by surprise by the revival of Vesuvius as they were visiting Pompeii. The town of Cercola came out of the exercise best, but dozens of people from Pompeii and Ercolano (Resina) did not complete the course because heavy rain had slowed down access to the autostrada.

However, even if the evacuation could be carried out, the experts still have to forecast the date of the eruption accurately. They cannot afford to err. Therefore, they would have to organize precautionary evacuations. However, the administrative problems would not end there. Refugees would not be grateful and docile for long if they were whisked from their homes too early, and unrest would soon develop if the emergency outlasted their patience. The refugees will not relish the prospect of sheltering indefinitely in distant tents, schools or hostels: they would prefer to be at home, even if that home were in danger of destruction. On the other hand, if the people are evacuated too late, the experts and authorities might then be blamed for the indescribable panic and thousands of deaths, and surviving relatives would certainly complain.

The most recent evacuation approaching this scale took place in June 1991, when 58 000 tribal people were removed from the forests as the Philippine volcano, Pinatubo, was erupting. The evacuation of over ten times that number from the congested urban environment around Vesuvius would be of a greater order of magnitude altogether, and on a tight schedule with

**Alert levels for an eruption of Vesuvius and the civil action proposed
(Dipartimento della Protezione Civile)**

Alert level	Alert state	State of volcano	Proposed action
0 No alert	Low	Typical background values	None
1 Caution	Medium	One monitored indicator departs from background values	Experts meet; population is alerted
2 Caution	High	One indicator departs from background values, suggesting a possible pre-eruptive state	Local administration provides logistic support for scientists
3 Pre-alarm	Very high	More than one monitored indicator departs from background values, suggesting a possible pre-eruptive state	Control passes to national government, which declares a state of emergency; civil-defence model for intervention is activated
4 Alarm	Maximum	Several indicators all point to a pre-eruptive state; eruption expected within a few weeks at most	Zona Rossa is completely evacuated; refugees taken outside Campania by car, bus, or boat to the twin regions
5 Waiting	Maximum	Data show pre-eruptive conditions; volcanic situation probably irreversible	Civil-defence operators and scientists leave the Zona Rossa
6 Eruption	Maximum	Eruption in progress	The people in part of the Zona Gialla suffering heavy ashfalls are evacuated to places within Campania
7 Aftermath	Maximum	Eruption ends; attention paid to late features such as mud-flows (notably in the Zona Blu) and gas emissions	Administration defines methods for population to return home; state of emergency is repealed

little latitude for manoeuvre. Thus, to make the most of the warning, the decision about whether to evacuate or not would have to be taken with some speed. The evacuation may also provide a robbers' paradise, as increasing numbers of houses would be left unoccupied, unprotected and impossible to guard, even in an area where heavy domestic protection is the norm. The revised plan endeavoured to counteract such criminality by proposing that the Zona Rossa should be encircled with a large police force. It is not to impugn the honour of the police to wonder whether this proposal would be physically possible, or at all feasible, especially since their devotion to duty might understandably wane when they realize that they themselves may be unable to escape when the eruption occurs.

The municipalities in the Zona Rossa and the reception regions with which they are twinned

Municipalities (with their populations)	Twinned receiving regions and means of transport to them
San Giorgio a Cremano (60 173)	Lazio (boat)
Portici (61 822)	Emilia Romagna (train)
Ercolano (Resina) (57 983)	Toscana (train)
San Sebastiano (10 320)	Molise (bus)
Pollena Trocchia, Massa di Somma (19 328)	Umbria (bus)
Ottaviano (24 728)	Piemonte and Valle d'Aosta (train)
Sant'Anastasia (28 888)	Marche (bus)
Somma–Vesuviana (33 673)	Abruzzo (bus)
Cercola (19 486)	Friuli & Venezia Giulia (bus)
San Giuseppe (26 820)	Lombardia (train)
Terzigno (15 954)	Veneto (train)
Boscoreale (29 305)	Apulia (train)
Pompeii (26 019)	Liguria (boat)
Torre del Greco (95 243)	Sicilia (train and boat)
Torre Annunziata (47 360)	Calabria (train and boat)
Trecase & Boscotrecase (20 005)	Basilicata (train and bus)

Communications and public awareness

In theory, once the warning signs have been registered, the experts would have time to warn the administrators, the media and the threatened population that an eruption is imminent. In theory, too, the evacuation plan already established would be put into operation, and the people in danger would be taken to safe places of refuge. In theory, the population will have been told what to expect, where to assemble, and where they are going.

In theory, the theory will work. Nevertheless, problems of intercommunication will still meet major barriers. In a volcanic crisis, Earth scientists, administrators, and journalists are locked together in a situation where efficient intercommunication is quite literally vital. But the specialists, who have to give the explanations, and the general public, who do not understand the workings of volcanoes, find themselves on different sides of a large barrier because they do not speak the same language. They are not trained in the same fields, they do not use the same jargon, and they do not have the same agenda. Unfortunately, Earth scientists are no better at producing soundbites and headlines than journalists are at studying volcanoes. Thus, on the one hand, Earth scientists seem to speak gobbledegook; and, on the other, the media seem to be full of misconceptions. At the same time, a volcanic crisis brings the role of administrators into a sharp focus. Administrators seldom see it as their prime duty to save any section of the community. An evacuation would probably cause as many problems for them as an eruption. They also know that they will be blamed for any miscalculations. Thus, the three groups are rarely seen to be pulling in the same direction and a volcanic emergency would magnify the chances of misunderstandings, misconceptions and mis-interpretations of events. Even more misunderstandings will probably arise if the special local tax, mooted in February 2007, is implemented to pay the extra costs of putting the contingency plans into effect.

The special problems of Campania

The task of evacuating no less than 600 000 people living in 18 cramped and congested towns before the eruption starts may well prove to be insuperable. The built-up areas extend, virtually unbroken, all along the northern shores of the Bay of Naples, and some of the most vulnerable sectors around Vesu-vius coincide with some of the most densely populated districts in Europe. The density of population in the area as a whole is about 3750 per km^2, but it reaches a maximum of over 17 000 per km^2 in Portici and San Giorgio a Cremano, and 14 000 per km^2 in Torre del Greco.

The population in this area has doubled since Vesuvius last erupted in 1944. Many people have been lured from the countryside by government-planned industrialization and employment schemes, and they have been frequently lodged in cheap, ugly, insalubrious and often illegally constructed blocks that amount to some of the worst housing conditions in Europe. These blocks were built without the protection of the established building norms, although it is claimed that further safeguards have been imposed upon constructions built since 1993. Such structures would not easily survive a powerful assault from Vesuvius.

In addition, one of the most remarkable and barely comprehensible fea-tures of the postwar period has been the way in which new buildings have been allowed to proliferate quite illegally higher and higher up the volcano in recent decades. Even apart from any considerations of common sense or environmental protection, it is a mystery to outsiders how these homes could possibly have been built in such a vulnerable situation. Other customs clearly rule on Vesuvius. The constructors of these illegal buildings then seem to have obtained retrospective *condoni* (permits that legalize these peccadilloes). One result is that those who have built, sold or occupied these dwellings have a vested interest in believing, and fostering the conviction in others, that

Vesuvius is in fact extinct and presents no danger to those who want to live on its flanks and enjoy the splendid view of the bay.

Once the experts judge that an eruption seems inevitable, it is envisaged that 16 500 police and soldiers would organize the evacuation of 80 000 people each day, using a flotilla of 81 ships, 40 trains from the Circumvesuviana railway, and 4000 cars. An evacuation on such a scale would be a logistical nightmare even during a period of calm. If a vision of the Last Judgment in such a huge emergency is to be avoided, an evacuation would require meticulous organization to a level that has scarcely ever been seen in Europe, and also obedience and discipline on the part of the local population, which they have not displayed since the Roman empire was at its most efficient. And this is assuming that the harbours and railways, and the labyrinthine network of narrow crowded roads, were not completely blocked by all those who had decided to ignore the public plans and make their own travel arrangements. However, critics have pointed out that no clear network of escape routes has been designated, and that the roads are so congested at present that nothing more serious than a rainstorm can paralyse public transport (as indeed occurred during the Mesimex exercise in 2006). The system therefore could fail spectacularly after Vesuvius emits its warning earthquakes.

Naples is no ordinary city and Campania no ordinary province. The prospects of the next eruption of Vesuvius offer a truly Neapolitan scenario, full

Housing encroaching into the danger zone on the southern slopes of Vesuvius above Torre del Greco (courtesy of David Alexander).

of polemics, passions, frustration, fatalism, indifference and, perhaps inevitably, chaos. The people resent injunctions from the legitimate government, a resentment ingrained in their consciousness after almost a millennium of rule by foreign dynasties. The Neapolitan region suffers from notorious social handicaps ranging from corruption, clientelism and nepotism, property speculation, unemployment, poverty, illiteracy, and, not least, a parallel government. Some assert that the administration is corrupt, undemocratic and unable to organize its own affairs, let alone able to combat the parallel government. An eruption would be just another scourge to add to the list. Many people nowadays say that their present unemployment worries them far more than the distant prospects of an eruption. Some might call this apathy, but it is more likely that the people accept their fate with the resignation of those well used to coping with a harsh and unforgiving world.

The ordinary people, the victims of this panoply of scourges, are practically powerless to do anything about them. After centuries of experience, they also have a deep and justified mistrust of any administrator, and an equally justified fear of organized crime. The poor, too, have few transferable assets, and many would prefer to stay at home and guard what little they have, as they seldom have anywhere else to go. The richer members of the community would almost certainly grasp their chequebooks and credit cards, pack their jewellery, jump into their cars and drive off to safety, probably blocking the escape roads in the process. On the other hand, when the emergency is declared, it is very difficult to imagine people in the Zona Rossa or Zona Gialla queuing patiently with their luggage in the reception centres, waiting for the train to take them to Calabria.

Specialists in matters of public psychology say that people living in threatened areas are blind to the danger and they are totally confused about what might happen to them and what they should do about it. The Campanians might agree that the disaster is certain to happen somewhere on the flanks of Vesuvius, but they may well refuse to believe that this disaster will occur where *they* live. When the crisis comes, they will be playing Russian roulette with the volcano, and it is not a game that volcanoes usually lose.

Popular attitudes to a volcanic disaster extend back into the mists of time. Vesuvius is also one of those volcanoes that is a mascot, a talisman and an emblem to all those who live around it. It has always been on the horizon, watching over the Campanians, almost like some enormous lucky charm. They regard Vesuvius with affection until it sends ash or lava into their homes. Even then, they are seldom afraid of it, or angry with it for very long.

Thus, the people living around a volcano even as ferocious as Vesuvius have ambiguous and conflicting attitudes towards it.

It is not surprising, then, that the contingency plans to deal with the next eruption have provoked acrimonious arguments among the volcanological experts, and sometimes also to a collective shrugging of the shoulders from those who will surely figure among the victims of the catastrophe. Many of the inhabitants are poorly educated, and many, too, have a fatalistic attitude to the volcano: "What", they ask, "can they do about it?", and "Nothing" is their almost invariable answer. There are also many who remain far from persuaded that Vesuvius presents any danger to them whatsoever. Other citizens are convinced that they will have to ignore the directives of any plans, use their initiative and fend for themselves, as usual, without having any clear idea about what they should do.

It was partly to counteract these attitudes that the scientists working on the Exploris Project, devoted to Vesuvius and other dangerous volcanoes, have recently concentrated upon correcting these fatalistic notions.

There is a grave danger, too, that many ordinary people will put their trust in the certitudes of unscrupulous and uninformed pundits, as they did, for example, in Colombia when Nevado del Ruiz erupted in 1985. However, Naples being Naples, there may be an even greater chance that they will follow a religious route. The Campanians would probably place more trust in ministers of religion than in ministers of government, and many of the threatened people would perhaps place even greater trust in a prayer to San Gennaro than in any administrative circular about a forthcoming eruption. After all, Neapolitans are perfectly aware that San Gennaro has protected the city for 1500 years, and they are equally aware that successive governments have offered no such succour. Other settlements around Vesuvius have also been protected, but less successfully on the whole, by their own particular saints, and their citizens would no doubt behave in the same way. Nevertheless, although appeals to the saints provide comfort and consolation, they also have the effect of fostering public inertia, which will not be the required reaction when the next eruption is imminent.

Some counter-suggestions

Inevitably, the plan to deal with the forthcoming volcanic emergency met with strong criticism, perhaps most controversially from Flavio Dobran, who particularly emphasized its social and cultural inadequacies. He and other critics stressed that it does not take into account the special needs, attitudes, emotive reactions, and conditions and traditions of those under threat. In 1995, Flavio Dobran initiated a multidisciplinary programme, Vesuvius 2000, and set up a group aiming to avoid a wholesale evacuation by finding ways of making it as selective as possible. They wanted to make the threatened area safer by studying the volcano in all possible aspects, and to educate the ordinary people so that they would properly understand the problems involved, know how to live with the threat, and know exactly how to react without panic when the inevitable crisis came. To this end, many lectures, seminars, workshops and discussion groups within educational, administrative and cultural organizations have been undertaken, and books, pamphlets and newsletters have increased public awareness of the danger.

Flavio Dobran then developed a global volcanic simulator specifically with the aim of generating a computer model that would provide the most accurate and all-embracing forecast of the future behaviour of Vesuvius. Again, he used a multidisciplinary approach, and brought together and quantified masses of geological and geophysical data ranging from the magma supply to the nature of the volcanic vent, and to the way in which the erupted materials would spread after they left the crater. The simulation confirmed previous indications from statistical studies that the next eruption could be expected during the twenty-first century, that it would be at least as powerful as that in 1631, but that the warning time would probably be limited to about a week rather than the two or three weeks indicated by the commission. The simulator suggested that, if the eruption were to be as powerful as that in AD 79, Vesuvius could generate the slowest type of pyroclastic flow that could travel the 7 km to the Bay of Naples in five minutes (at 84 km an hour). If the eruption were as powerful as that in 1631, a similar flow would still take only ten minutes to reach the bay. These speed figures seem to be conservative estimates, although pyroclastic flows travelling at such slow speeds would nevertheless allow very little time for individuals to escape.

The global volcanic simulator indicated that pyroclastic flows from an eruption as powerful as that in 1631 could be slowed down and halted by two well built barriers, 30 m high, placed at distances of 2.5 km and 4.5 km respectively from the crater. It may not be even necessary to build these barriers, because, for instance, a similar role could be probably be filled by the existing solid blocks of buildings, 25 m high, which have been properly constructed according to the legal building requirements. These alternative suggestions at least serve to demonstrate that it is possible to envisage different solutions to this tantalizingly difficult environmental problem.

Relocation

In the past, whenever the dust has settled and the lava has cooled, the Campanians have usually set about rebuilding their homes in virtually the same places as before. In the latest instance, San Sebastiano was rebuilt over the lavas that destroyed parts of the old town in 1944. Torre del Greco was partly destroyed in 1631, 1737, 1794 and 1861, and each time the citizens reconstructed their town over the ruins. In 1794, King Ferdinand offered the people a new site near by, where they could rebuild their homes. They refused. Perhaps they had fallen under the spell of the *genius loci*, the spirit of the place; perhaps they hoped, prayed or believed that Vesuvius would send its destructive emissaries to some other town next time. Those who understand and admire this persistence call it resilience, fatalism or stoicism; those who do not understand may call it insouciance.

One method of alleviating the problems could be by reducing the threatened population in the danger zones by offering the residents special incentives to leave. In July 2003, Marco di Lello, the regional urban assessor, proposed the VesuVia decongestion plan, to reduce the population in the Zona Rossa by 20 per cent. This was a long-term project, aimed at helping poorer families living in rented accommodation, who earned less than 30 000 euro a year. Subject to various assessments, a *bonus casa* (cash incentive) was offered in the hope of persuading 150 000 tenants to start a new life beyond the threat of an eruption. Each suitable family was then offered a maximum of 25 000 euro (later raised to 30 000 euro) towards a down payment to buy a house outside the Zona Rossa. The offer was open to Italian families, as well as to foreigners who had been resident in the area for five years, a requirement that was subsequently reduced to three years in 2005. The owners of some houses in the danger areas could also sell their property to a specially established regional company. This decongestion plan was linked to measures for economic and ecological development. The homes left by their tenants, or sold by their owners, could then be destroyed or changed, for instance, into small museums, artisan workshops, or bed & breakfast houses, which would attract tourists and inject cash and diversity into the local economy. Such establishments, with few permanent residents, could be evacuated much more easily than whole families and their possessions.

In the first tranche in 2003–2004, 3276 families took up the offer, followed by another 2421 families in the second tranche in 2004–2005. The total of 5697 families would amount to about 25 000 people. Most requests have so far come from some of the towns that would probably be in the greatest danger during the next eruption: Torre del Greco led with 876 requests in the two tranches, Portici registered 762, San Giorgio a Cremano 713, and Somma–Vesuviana 408.

Although the authorities remain optimistic, it seems that the offer has not yet met with universal success, and it is the older people – those who could remember the last time – who have apparently been most inclined to take it up. Perhaps the maximum sum of 30 000 euro was deemed too small; but it is probably more likely that the residents did not feel sufficiently threatened to move away. There have also been conflicting administrative signals. On 23 September 2004, for instance, the newspaper *Il Mattino* pointed out a fine example of administrative confusion in Torre del Greco. People were being given financial incentives to leave the Zona Rossa, while young couples were being offered grants to buy their first house in the very same area. Furthermore, hundreds of new inhabitants have settled in the Zona Rossa every year since the decongestion plan was implemented.

The possible criticisms of the present contingency plan, its applications and its alternatives are manifold. Many of these criticisms could be raised about any such emergency plan, but, by the admission of many Neapolitan experts themselves, the problems are exacerbated in the Campanian environment. Hence there is a need to persuade the people of the merits of emergency planning as well as the need for continual revision of the present contingency plan in the light of ever-increasing understanding of the complex processes of volcanic activity.

Conclusion

In the days before volcanology developed, eruptions were commonly seen in two contrasting ways. On the one hand, they were believed to be portents of military defeats or the deaths of great prelates or potentates. At the same time, however, eruptions were also seen as God's way of punishing sinners, which even the local patron saints such as San Gennaro found hard to alleviate. In Naples, in particular, prayers, penitence and penance were displayed in frenzied processions that probably had greater psychological value for the sinners than any practical impact on the effects of the eruption.

For most of the past 2000 years, the people living around Vesuvius have maintained a complex relationship with the volcano, in which fear, awe and admiration have predominated. Eruptions came as a complete surprise. The signs warning of an eruption went unknown or unheeded. There could be no forward planning. Palliative action could be undertaken only on the spur of the moment, usually on individual initiatives, and commonly after the eruption had all but ceased. In Campania, the greatest catastrophes produced the largest reactions. Thus, the eruptions in AD 79, 1631, and perhaps in c. 472, seem to have witnessed the most humanitarian aid.

The persistent but less powerful eruptions after 1631 had a more limited regional impact that called for much less aid and regeneration. For many Campanians, these eruptions were more of a sublime spectacle than a threat, because they damaged relatively small areas, although they often produced a wonderful show that greatly encouraged the tourist trade.

Then, from the middle of the eighteenth century, Earth scientists began to break the codes that determined the behaviour of volcanoes, and the students of Vesuvius were in the forefront of this research, especially after the world's first volcano observatory was opened on its flanks in 1845. Vesuvius was near a flourishing centre, it was in nearly constant activity, and it was watched by experts, who thereby refined their understanding. Eventually, what for long had seemed to be random spasms of nature could be interpreted as signs of an imminent eruption. If the signs were correctly interpreted, volcanology could become a predictive science and eruptions could be forecast; contingency plans could be made; lives could be saved.

However, the contingency plans come up against attitudes and beliefs that have become traditions engrained over the centuries. It is vital that these attitudes should change, because the next eruption is likely to be the most powerful since 1631. The longer Vesuvius remains dormant, the more violent will be its revival. Vesuvius is not extinct. It is not harmless. It is not a talisman for Campania. It is not merely the icon of a benign picture postcard, dominating a beautiful bay. Future generations must accept that they will have to deal with a really dangerous volcano, one of the most dangerous in the world. The better the Campanians are prepared to face this threat, the more likely they will be to survive and live to tell a remarkable tale.

Contingency planning to cope with the next eruption is vital; there is a greater need for this than ever before. In recent decades, the population around Vesuvius has not only multiplied in numbers and density, it has also encroached more and more into the volcanic domain. Thus more people are now vulnerable to the threat of an eruption of Vesuvius than at any time in its whole history. If, for instance, the next eruption kills the same proportion of the population as died in AD 79, it would cause an unprecedented volcanic catastrophe. Thus, the contingency plan to deal with such an emergency has to be as fool proof as possible; it must be implemented with great efficiency and it cannot be allowed to fail. Otherwise over half a million people could be in grave danger of succumbing to a horrible death.

Further reading

Il Mattino; La Repubblica Napoli; Balmuth et al. 2005; Barberi et al. 1983; Chester 1993; Chester et al. 2002; Dipartimento della Protezione Civile 1995 et seq.; Dobran 1993, 1994, 1995, 1998, 2006; Macedonio et al. 1990.

APPENDIX 1

The two letters of Pliny the Younger to Tacitus about the eruption of AD 79

(translated by H. M. Hine, Scotstarvit Professor of Humanity, University of St Andrews, Scotland)

The eruption that killed Pliny the Elder (Book 6, Letter 16)

Thank you for asking me to send you a description of my uncle's death so that you can give a more accurate account of it to posterity. I know that immortal fame awaits him if his death is recorded by you. It is true that he perished in a catastrophe that destroyed the most beautiful regions of the Earth, a fate shared by whole cities and their populations, and one so memorable that it is likely to make his name live for ever. He himself wrote several books of lasting value, but you write for all time and can still do much to perpetuate his memory. In my opinion, the fortunate man is one to whom the gods have granted the power either to do something that is worth recording or to write something that is worth reading, and the most fortunate of all is the man who can do both. My uncle will be such a man, thanks to his writings and yours. So you set me a task that I would choose for myself, and I am more than willing to start upon it.

My uncle was stationed at Misenum in command of the fleet. On 24 August, at about one o'clock in the afternoon, my mother pointed out to him a cloud that could be seen which had an odd size and appearance. He had been out in the sunshine, had taken a cold bath, and had lunched while he was lying down, and was then working at his books. He called for his shoes and climbed up to a place that would give him the best view of the amazing phenomenon. From that distance it was not clear from which mountain the cloud was rising – although it was afterwards found to be Vesuvius. The cloud could best be described as more like an umbrella pine than any other tree, because it rose high up in a kind of tall trunk and then divided into branches. I imagine that this was because it was thrust up by a recent blast until its power weakened and it was left unsupported, or was even overwhelmed by its own weight and spread out sideways. Sometimes it looked pale, sometimes it looked mottled and dirty, according to whether it had carried up earth or ash. Like a true scholar, my uncle saw at once that it deserved closer study, and he ordered a boat to be prepared. He said that I could go with him if I wished. I replied that I would rather continue with my studies. As it happened, he himself had given me some writing to do.

Just as he was leaving the house, he was handed a message from Rectina, the wife of Tascus, whose home was at the foot of the mountain, and she had no way of escape except by boat. She was terrified by the threatening danger and begged him to rescue her from disaster. He changed plan at once; and what he had started in a spirit of scientific curiosity, he ended as a hero. He ordered the large galleys to be launched and went on board ship himself. He intended to bring help to many other people as well as Rectina, because this beautiful stretch of the coast was densely populated. He hurried towards the place from which everyone else was fleeing and steered a straight course right into the danger, so fearlessly that he dictated notes on all the changes that he observed as the disaster unfolded.

The ash that was already falling became hotter and thicker as the ships approached the coast. Soon there was pumice and blackened, burnt stones that had been shattered by the fire. Suddenly the sea shallowed where the shore was obstructed and choked by debris from the mountain. For a moment, my uncle wondered whether to turn back, as the captain advised, but he decided instead to go on. "Fortune favours the brave", he said, "take me to Pomponianus". Pomponianus lived at Stabiae and was separated from the boat by the width of the Bay of Naples because the shore gradually curves around a basin filled by the sea. Pomponianus was not yet in danger, but he would clearly be threatened as the danger spread. In fact, Pomponianus had already put his belongings into a boat so that he could escape as soon as the contrary, onshore wind changed. This wind, of course, was fully in my uncle's favour and quickly brought his boat to Stabiae. My uncle embraced his terrified friend and offered him comfort and encouragement. He believed that he could best calm his friend's fears by demonstrating his own composure, and thus gave instructions that he was to be taken to the bathroom. After his bath, he lay down and dined. My uncle was quite cheerful, or at least he pretended to be, which was just as brave.

Meanwhile, very broad flames, and fires leaping high, blazed out from several places on Vesuvius and the darkness of the night seemed to make them glare out with even greater brilliance. My uncle soothed the fears of his companions by saying repeatedly that they were nothing more than fires left by the terrified peasants, or empty abandoned houses that were blazing. He went to bed and fell properly asleep; because, as he was a stout man, his breathing was loud and heavy and could be heard by those passing his door. But, eventually, the courtyard that gave access to his room began to fill with so much ash and pumice that, if he had stayed in his bedroom, he would never have been able to get out. He was awakened and he joined Pomponianus and his servants, who had sat up all night. They wondered whether to stay indoors, or go out and take their chance in the open, because the buildings were shaking with frequent violent tremors and appeared to be swaying backwards and forwards as though uprooted from their foundations. Outside, there was the danger from the falling pumice, although it was only light and porous. After weighing up the risks, they opted for the open country. For my uncle, it was a choice between rational considerations, for the others it was a choice between fears. To protect themselves from falling objects, they tied pillows over their heads with cloths.

It was daylight everywhere else by this time, but they were still enveloped in a darkness that was blacker and denser than any night, relieved only by many torches and various other lights. The decision was taken to go down to the shore to see at close quarters if there was any chance of escape by sea, but the current was still dangerous and running in to the shore. He lay down to rest on a sheet that had been spread on the ground for him and he called once or twice for a drink of cold water and gulped it down. Then, suddenly, flames and a strong smell of sulphur that warned of yet more flames to come, forced the others to flee and get him up. He stood up, with the support of two slaves, and then he suddenly collapsed. I imagine that it was because he was suffocated when the dense fumes choked him by blocking his windpipe, which was constitutionally weak, narrow and often inflamed. When light returned on the third day after the last day that he had seen [on 26 August], his body was found intact and uninjured, still fully clothed and looking more like a man asleep than dead.

Meanwhile, my mother and I were at Misenum, but this is of no historical interest and you only wished to hear about my uncle's death. I will say no more except to add that I have described in detail every incident that I either witnessed myself, or heard about immediately afterwards, when the reports were most likely to be accurate. It is for you yourself to select what suits your purpose best, for there is a great difference between a letter and history, between writing for a friend and writing for everyone. *Vale*.

Pliny the Younger at Misenum (Book 6, Letter 20)

Thus, the letter that you asked me to write about my uncle's death has made you keen to hear about the terrors and the hazards that I myself had to face back in Misenum, because I broke off at the beginning of that part of my story. "Although my mind shrinks from remembering . . . I will begin" (*Aeneid* 2.12).

After my uncle left, I spent the rest of the day at my books, because this had been the reason I had stayed behind. Then I took a bath, dined and went to bed, but I slept only fitfully and briefly. We had experienced earth tremors for several days, which were not especially alarming in themselves because they happen so often in Campania. However, that night they were so violent that everything felt as if it were not only being shaken but also destroyed. My mother came hurrying to my room and found me already getting up to wake her if she had still been asleep. We sat together in the forecourt between the buildings and the sea near by. I do not know whether I should call this courage or foolishness on my part (I was only 17 at the time), but I asked for a volume of Livy and went on reading as if I had nothing else to do. I even continued with the extracts that I had been making. A friend of my uncle's had just arrived from Spain to join him. When he came up and saw us sitting there, and me reading indeed, he scolded both of us – me for my folly, and my mother for permitting it. Nevertheless, I remained intent upon my book.

By now it was past six o'clock and the dawn light was still only dim and feeble.

The buildings around were already tottering and we would have been in immediate danger if our house had collapsed, because we were in such a small and confined space This made us decide to leave town. We were followed and pressed ever forward by a dense and panic-stricken crowd that chose to follow someone else's judgement rather than decide anything for themselves – a point when fear looks like prudence. We stopped once we were beyond the buildings of the town, when some extraordinary and most alarming things happened. Although the ground was flat, the carriages that we had ordered began to lurch to and fro and we could not keep them still – even when we wedged their wheels with stones. Then we saw the sea sucked away and then apparently forced back by an earthquake. Certainly, the shoreline had retreated and many sea creatures were left stranded on the dry sand. From the other direction, a dreadful black cloud, torn by whirling quivering bursts of fiery gas, parted to reveal long tongues of flames like lightning, but larger. At this point, my uncle's friend from Spain spoke out even more urgently. "If your brother, if your uncle is still alive, he would want both of you to be saved. If he is dead, he wanted both of you to survive him – so why postpone your escape?" We replied that we would not think of considering our own safety as long as we could not be certain about his. Without further ado, our friend rushed off and hurried out of danger as fast as he could.

The cloud sank down onto the earth soon afterwards and covered the sea. It had already surrounded and hidden Capri from sight and had concealed the projecting part of Misenum. My mother begged, implored, and then commanded me to leave her and escape as best I could. A young man might be able to escape, whereas she was old and slow. She would be able to die in peace provided that she had not been the cause of my death as well. I told her that I refused to save myself without her. I took her hand and made her hurry along with me. She agreed reluctantly and blamed herself for slowing me down. Ash was already falling by now, but not yet very thickly. Then I turned around: a thick black cloud threatened us from behind, pursuing us and spreading over the land like a flood. "Let us leave the road while we can still see", I said, "or we will be knocked down and trampled by the crowd in the darkness". We had hardly sat down to rest when the darkness spread over us. But it was not the darkness of a moonless or cloudy night, but just as if the lamp had been put out in a completely closed room.

You could have heard women shrieking, children crying and men shouting. Some were calling for their parents, some for their children, some for their wives, and trying to recognize them by their voices. These people were bewailing their own fate, or those of their relatives. Some people were so frightened of dying that they actually prayed for death. Many begged for the help of the gods, but even more imagined that there were no gods left and that the last eternal night had fallen on the world. There were also those who added to our real perils by inventing fictitious dangers. Some claimed that part of Misenum had collapsed, or that another part was on fire. It was untrue, but they could always find somebody to believe them.

A glimmer of light returned, but we took this to be a warning of approaching fire rather than daylight. But the fires stayed some distance away. The darkness came back and ash began to fall again, this time in heavier showers. We had to get up from time to time to shake it off, or we would have been crushed and buried under its weight. I could boast that I never expressed any groan or cry of fear in this time of peril, but I was kept going only by the thought that I was perishing together with everyone else, and that everything was perishing with me, a wretched, but considerable consolation for my mortal condition.

After a while, the darkness paled into smoke or mist, and the real daylight returned, but the sun shone as wanly as it does during an eclipse. We were terrified by what we saw, because everything had changed and was buried deep in ash like snow. We went back to Misenum, where we attended to our physical needs as best we could, and then spent an anxious night alternating between hope and fear. Fear was uppermost because the earth tremors were still continuing and several hysterical people made their own and other people's misfortunes seem trivial with their terrifying predictions. But even then, in spite of the dangers we had undergone and were still expecting, my mother and I still had no intention of leaving until we received news of my uncle.

Of course, these details are not worthy of history, and you will read them without wishing to record them. If they seem hardly worth even putting in a letter, then you have only yourself to blame for asking for them. *Vale.*

APPENDIX 2

Cassiodorus: Variae Epistolae, letter 50

*The letter about a recent eruption of Vesuvius, written by Cassiodorus
on behalf of King Theodoric, the ruler of Italy, to Faustus,
his Pretorian Prefect of Campania*
(translated by H. M. Hine, Scotstarvit Professor of Humanity,
University of St Andrews, Scotland)

The Campanians, devastated by the hostility of Mount Vesuvius, and having been
stripped of the fruits of their fields, have poured forth tears and implored our clemency, so that they may be relieved of the burden of paying tax. Our piety rightly
assents to this.

But, because the misfortune of each individual is unclear to us unless it is examined, we instruct your Greatness [Faustus] to send a man of proven reliability to the
territory of Nola or Naples, where poverty itself is advancing because of the damage
to property. In that way, once the land there has been carefully inspected, relief may
be given in accordance with the damage to the owner's livelihood. When the whole
extent of the damage is recognized, let the amount of benefit be accurately matched
with it.

For the province, which has been deflowered of its lands, labours under this one
evil, that, lest it enjoy perfect happiness, it is frequently shaken by the cruelty of this
fearsome phenomenon. But that terrible event is not totally unfeeling: for it sends
significant warning signs, so that the adversity may be borne with greater fortitude.

The opening on the mountain murmurs when nature is struggling with such great
masses [within it], so that the wind that is stirred up terrifies the neighbourhood with
resounding roaring. The air there is darkened by the most foul exhalation, and, when
that destruction is set in motion, it is recognized virtually throughout the whole of Italy.
Burnt-out dust flies through the great void, and when clouds of earth have been stirred
up, the dusty drops rain down on the provinces overseas as well, and what Campania
can be experiencing is recognized when its suffering is felt in another region of the
Earth.

You would see, as it were, some rivers of dust in motion, and, as it were, liquid
currents rush down in a sterile seething mass. You would be amazed that the surface
of the fields had suddenly swollen up as high as the tree tops; and what had been coloured the most fertile green, was suddenly devastated by the calamitous heat. That
perpetual furnace spews out sand formed from pumice, but it is still fertile. Although
the sand has been dried up by lengthy burning, it soon puts forth shoots that develop
into various plants, so that what had been devastated shortly before is restored with
great speed. What is this extraordinary anomaly whereby one mountain should rage

in such a way that it is proved to terrify so many sections of the world . . . and to scatter its substance everywhere, without seeming to suffer any loss?

It showers down dust far and wide, and belches out great lumps on the local people, and has been regarded as a mountain for so many generations, although it is consumed by such great emissions. Who would believe that such huge chunks have burst out and have been carried from such deep caverns right down to the plains; and that, as it were, light chaff has been thrown out, spewed from the mountain's mouth by an exhaling wind?

In other parts of the Earth great peaks are seen to burn locally: virtually the whole universe has been granted knowledge of the fires of this mountain. How then could we not believe the inhabitants in a matter that can be recognized by the testimony of the universe? Therefore, as has been said, let your prudence choose such a man as can both bring relief to those who have suffered harm and leave no room for fraud.

Glossary

ash Pulverized volcanic rock or magma exploded violently from a vent in fragments less than 2 mm in size. It can form widespread blankets on land and forms extensive layers covering Campania.

ashflow A turbulent mixture of volcanic ash and gas, which is expelled from a vent at high temperatures and at great speed from a vent and covers vast areas like a large pyroclastic flow. The Campanian Ignimbrite and the Neapolitan Yellow Tuff were probably expelled in this way.

basalt A dark volcanic rock, with only about 40–52 per cent of silica, but relatively rich in iron, calcium and magnesium. By far the most common volcanic rock, forming much of the ocean floors and most of the lava flows on land. It usually emerges hot and fluid (1100°C–1200°C) onto the land surface, and flows are commonly 10 km long. It usually erupts without violent explosions, from fissures as well as single vents.

benchmark A clearly defined point on the land surface whose height has been accurately determined by precise geodetic levelling.

bradyseism A term derived from the Greek, *bradus*, slow, and *seismos*, movement. Small oft-repeated movements of the Earth's crust that bring about slow vertical displacements of the land surface. They are often caused by oscillations in the levels of subterranean magma. Many such crustal movements that have taken place in the Campi Flegrei during the past 10 000 years.

caldera A steep-sided almost circular or horseshoe shape hollow, several kilometres across, formed on a large volcano by violent explosions or by the collapse of the summit. They are usually bounded by steep enclosing walls and formed most often on large volcanoes. The term is derived from the Spanish word for cauldron. Much of the Campi Flegrei lie within a caldera. The remains of the old caldera of Somma–Vesuvius now form the ridge of Monte Somma.

Campanian Ignimbrite See ignimbrite.

cinders Fragments of lava, commonly between 64 mm and 30 cm in size, that are expelled by moderate explosions, and often form cones up to 250 m high. They are usually light, rough, and riddled with gas holes. They are also commonly known by their Italian name of *scoria*.

crust The solid outer layers of the Earth forming both the continents and the ocean floors. The crust is divided into plates.

dome A rounded dome-shape accumulation of lava that was too viscous to flow far from the vent.

effusion An eruption of molten lava which takes place with little, or no gaseous explosions and thus most commonly forms lava flows.

epicentre The point on the Earth's surface directly above the origin of an earthquake.

eruption The way in which gases, liquids and solids are expelled onto the Earth's surface by volcanic action ranging from violently explosions to noiseless effusive or hydrothermal emissions.

fissure A crack, fault, or cluster of joints, extending from the Earth's surface deep into the Earth's crust, which may allow magma to reach the Earth's surface; sometimes called a fracture.

fragments Ash, bombs, cinders, or pumice shattered by explosions during an eruption. They are the main constituents of many volcanoes. Also called pyroclasts and tephra.

fumarole A small vent giving off gases or steam, and often surrounded by fragile precipitated crystals. They are a major aspect of hydrothermal activity, along with solfataras and geysers.

geological time The whole history of the earth, extending back about 4.6 billion years.

historical time The timespan during which events have been recorded, in however fragmentary a fashion, by observers. In the Mediterranean area it may reach back 3000 years, whereas in the New World it can be less than 200 years.

hydrothermal eruption An eruption of gases, steam, and hot water, without magma; typically seen at La Solfatara.

hydrovolcanic eruption An explosive eruption made more violent by the addition of water derived, for example, from shallow sea-water, lakes, streams or aquifers. Hydrovolcanic activity played a predominant role in the formation of Monte Nuovo; and many eruptions of Vesuvius have had hydrovolcanic components. Also known as hydromagmatic eruptions.

ignimbrite Voluminous deposits, often covering more than $1 km^3$, of pumice, broken crystals, and glass in a matrix of ash, that are laid down by large pyroclastic flows or ashflows. The fragments can be welded together when they are deposited at high temperatures.

lava Molten rock or magma which reaches the surface, flows down slope and solidifies on cooling. Lava occurs as flows, domes, fragments within cones and as pillows formed on the ocean floors. Depending on their composition and temperature when they are emitted, some flows have markedly rugged surfaces; others are smoothly undulating; and yet others are blocky. The term seems to have been introduced from the Neapolitan word, "lava" ("wash") by Serao in his account of the eruption of Vesuvius in 1737. In volcanic terms, the word originally described the movement of the molten material, the flow, and then by extension was applied to the rock itself. In fact, to the Neapolitans, "lava" referred to any current that washed down from the heights. Thus, they used the same term for features of vastly different origins. They came to distinguish two sorts of these currents: *lava di fuoco* (literally "lava of fire", or what would now be called a "lava flow"); and *lava di acqua* (literally "lava of water", or what would now be called a "mudflow", or a "torrent"). Unfortunately, such distinctions were not refined enough to enable modern scientists correctly to interpret the course of many older eruptions. The classic case is the eruption of Vesuvius in 1631.

lava fountains Fountains made up of clots of molten lava. ranging in size from 1 cm to about 50 cm, often accompanied by ash, dust and fumes that are hurled up to 3–4 km into the air in spectacular jets. One of the most beautiful features of the volcanic repertoire.

leucite A white potassic feldspathoid crystal which is common in the lavas of Vesuvius.

levelling line An accurately surveyed line along which differences in height are precisely determined by levelling instruments.

magma Hot, mobile rock material, mainly formed by partial melting of the mantle, commonly at depths between 70 km and 200 km. It is composed of hot, viscous, liquid material containing hot, but still solid, crystals, or rock fragments and small proportions of included gases. It is less dense than the materials surrounding it and is thus able to rise slowly towards the Earth's surface by buoyancy. If it overcomes the pressure and resistance of the rocks of the Earth's crust, it erupts in a fluid state, releasing its contained gases with varying degrees of explosive violence and emits lava in flows or fragments. Magma temperatures can vary from 900°C to 1200 °C.when it is erupted onto the surface. The whole mass has rather the consistency of molten glass, but it has a much more complicated composition.

mudflow A current of water and a great proportion of fragments of all sizes and types that is commonly formed when an eruption melts part of an ice-cap, or disturbs a crater lake or when unconsolidated volcanic ash and pumice, for instance, is deluged with heavy rain. Mudflows travel down valley at high speed and often cause much damage. Often known by the Indonesian term, lahar.

Neopolitan Yellow Tuff See tuff.

phonolite A pale volcanic rock, rich in sodium and potassium, but relatively poor in silica (55–60 per cent), and usually emitted at temperatures of less than 1000°C.It is derived from alkaline basalts, and is so named because it breaks into plates that ring when struck.

plate The rigid slabs into which the lithosphere is divided. Their edges constantly diverge, or converge and plunge beneath each other. All are composed of oceanic crust and some also carry continental crust above it. Between 10 and 15 major plates, and a similar number of micro plates are generally recognized.

Plinian eruption A powerful explosion of gas, steam, ash, and pumice, lasting for several minutes, or hours, which rise in vertical columns to 30 km or more, and often branch out in the form of a Mediterranean pine, like that described by Pliny the Younger at Vesuvius in AD 79. Whenever the vertical impetus wanes, a part of the column might collapse and form pyroclastic flows. They are generated mainly by silicic magmas, although they can develop from hydrovolcanic eruptions of basalt. The gas and dust expelled to the stratosphere often form an acidic aerosol that can modify the weather over large tracts of the Earth for several years.

pumice Very pale volcanic fragments riddled with gas-holes, formed by the expansion of contained gases as the magma reaches the surface and then explodes very violently over vast areas during an eruption. Most pumice floats on water, and sometimes forms ephemeral floating islands after eruptions at sea. Its appearance varies from small fibrous chips to knobbly lumps and often resembles solidified foam. It is a major component of Plinian eruptions.

pyroclastic flow　An incandescent cloud, or glowing avalanche, of scorching hot gas and fragments of all sizes, including ash, pumice and rock debris in an aerosol-like emulsion expelled by explosive eruptions, which travels across the ground at very high speeds and gives off glowing billowing clouds. They form: after large blasts of gas; when large eruptive columns collapse and race down slope; or when the rims of craters collapse into the vent. They are perhaps the most dangerous of all the forms of volcanic eruptions. They are called pyroclastic surges when composed of a less dense cloud of ash and dust. They are also known as nuées ardentes, from the French for "burning cloud".

pyroclastic surge　See pyroclastic flow.

Richter scale　An open-ended scale devised by the American geologist, Charles Richter to measure the magnitude of earthquakes at their source. The weakest earthquakes, up to magnitude 3.0 are recorded but not felt by humans; those of magnitude 6.0–8.0 or more are very powerful. It is a logarithmic scale; thus an earthquake of magnitude 5.0 is ten times as powerful as one of magnitude 4.0. The Richter scale does not directly measure destruction at the surface. Such damage is measured on the 12-point scale of earthquake intensity first devised by Mercalli and later expanded and modified.

solfatara　An Italian word used to describe the emission of sulphurous gases from a fumarole, which is typical of the eruptions at La Solfatara in the Campi Flegrei.

Strombolian eruption　Repeated moderate effusive–explosive activity, expelling fragments of lava, that produces cones of ash, cinders and copious lava flows or lava fountains over a period of several months or a few years. The most common kind of eruption on land, it is typical of the activity of Stromboli in the Aeolian Islands, and was a prominent feature of much of the persistent activity on Vesuvius after 1631.

subduction zone　A zone where two plates converge and one plunges beneath the other into the mantle. The subducted slab releases volatiles that stimulate melting in the wedge of mantle above it that helps form volcanoes. The plunging action of the slab also generates deep-seated violent earthquakes.

sub-Plinian eruption　A less violent form of Plinian eruption, characterized by that of 1631 on Vesuvius.

trachyte　A pale, greyish volcanic rock relatively rich in silica (60–65 per cent) and in sodium and potassium, which is usually emitted at temperatures about 1000°C. It is viscous, forms domes and rugged lava flows, and is sometimes involved in violent explosions.

tsunami　A Japanese term used to describe large, fast-moving sea waves generated after violent eruptions or earthquakes that have taken place near or below sea-level. Scarcely detectable in the open sea, they increase in size and speed in shallow water, and crash onto the shore where they cause great damage and death.

tuff　Fine yellowish fragments exploded during eruptions often having a major hydrovolcanic component, which sometimes weld together on cooling. They a form thick layers in Campania, where they have often been used as building stones.

vent　The usually vertical conduit or chimney up which volcanic material travels from the magma source to the earth's surface. Sometimes also called a chimney, bocca, or mouth in older accounts.

volcanic explosivity index　An index of the violence of volcanic eruptions (VEI) devised in 1982 by Newhall and Self that enables the power of eruptions to be compared.

volcanic gas　The volatile component of magma, mainly including steam, carbon dioxide, sulphur dioxide, or hydrogen sulphide and smaller amounts of chlorine and fluorine. As the magma approaches the Earth's surface, the gases are released and can then become the chief factor making eruptions violent.

Bibliography

Original spellings are retained in antique sources, notably "Pompei".

Newspaper sources

Giornale del Regno delle Due Sicilie: daily newspaper published in Naples (notably 23–28 October 1822).

Roma (www.ilroma.net): daily newspaper published in Rome (notably 13 April 1906).

The Times (www.timesonline.co.uk): daily newspaper published in London (notably 22 March 1944).

Il Mattino (www.ilmattino.it): daily newspaper published in Naples. Recent articles include:

  02/01/04: Appello ai sindaci per il recupero degli antichi tracciati vesuviani

  01/03/04: Una sonda per ascoltare il respiro del Vesuvio

  11/03/04: Rischio Vesuvio, sarà allargata la zona rossa

  17/03/04: Marzo 1944, fuoco dal Vesuvio

  23/05/04: Esodo dal Vesuvio. Arrivano gli incentivi

  06/07/04: Sviluppo e ambiente le sfida del Vesuvio

  15/07/04: Arriva il bonus per chi lascia la zona Rischio-Vesuvio

  22/07/04: Due strade verso il volcano

  06/09/04: Fuoco e paura sul Vesuvio, fuga dalle case

  23/09/04: No al piano rischio Vesuvio

  26/10/04: Vesuvio; piu facile il bonus

  07/11/04: Osservatorio geofisico riapri la sedestorica

  19/12/04: Parco del Vesuvio

  12/09/06: Fuga-show, il rischio Vesuvio diventa evento

  18/10/06: Vesuvio: prove di fuga. Via al piano di emergenza

La Repubblica Napoli (napoli.repubblica.it): daily newspaper published in Naples. Recent articles include:

  18/10/06: Esercitazione sul Vesuvio fuga simulate per duemila

  21/10/06: In cento gouaches e dipinti storie del Vulcano ritrovato

  23/10/06: Il nubifragio rallenta le prove di evacuazione per il Vesuvio

  23/10/06: Scatta la sfida virtuale con il volcano ma la pioggia rallenta gli "evacuate"

  23/10/06: Vesuvio, prove di evacuazione Napolitano: non è innocuo

Il Denaro (www.denaro.it): daily newspaper published in Naples

  13/02/07 Vesuvio, si ipotizza una tassa di scopo per garantire più sicurezza

Acton, H. 1956. *The Bourbons of Naples, 1734–1825*. London: Methuen.

Adam, J. P. 1986. Observations techniques sur les suites du séisme de 62 à Pompéi. In Albore Livadie (1986b: 67–89).

Adams, F. D. 1990 (1954). *The birth and development of the geological sciences*. New York: Dover.

Addison, J. 1726. *Remarks on several parts of Italy in the years 1701, 1702, 1703*. London: Tonson.

Albore Livadie, C. 1986a. Considérations sur l'homme préhistorique et son environnement dans le territoire phlégréen. In Albore Livadie (1986b: 189–205).

———— (ed.) 1986b. *Tremblements de terre, éruptions volcaniques et vie des hommes dans la Campanie antique*. Publication VII (2nd series), Centre Jean Bérard, Institut Français, Naples.

———— 2002. A first Pompeii: the Early Bronze Age village of Nola-Croce del Papa (Palma Campania phase). *Antiquity* **76**, 941–2.

Albore Livadie, C., G. D'Alessio, G. Mastrolanzi, G. Rolandi 1986. Le eruzioni del Somma–Vesuvio in epoca protostorica. In Albore Livadie (1986b: 55–66).

Albore Livadie, C., L. Campajola, A. D'Onofrio, R. K. Moniot, V. Roca, M. Romano, F. Russo, F. Terrasi 1998. Evidence of the adverse impact of the "Avellino Pumices" eruption of Somma–Vesuvius on Old Bronze Age sites in the Campania region (southern Italy). *Quaternaire* **9**, 37–43.

Alexander, D. 1997. The study of natural disasters, 1977–1999: some reflections on a changing field of knowledge. *Disasters* **21**, 284–304.

Alfano, B. G. & I. Friedlander 1929. *La storia del Vesuvio illustrata dai documenti coevi*. Ulm: K. Hohn.

Amato di Montecassino *c.* 1178. *Storia dei Normanni*. In Gasparini & Musella (1991: 174).

Andersen, H. C. 1871. *The fairy tale of my life*. London: Paddington Press [reissued in 2000].

Andronico, D. & R. Cioni 2002. Contrasting styles of Mount Vesuvius activity in the period between the Avellino and Pompeian Plinian eruptions and some implications for assessment of future hazards. *Bulletin of Volcanology* **64**, 372–91.

Andronico, D., R. Cioni, P. Marianelli, R. Santacroce, A. Sbrana, R. Sulpizio 1998. Introduction to Somma–Vesuvius. In *Cities on volcanoes, international meeting*. G. Orsi, M. Di Vito, R. Isaia (eds), 14–25. Naples: Osservatorio Vesuviano.

Ayala, Simon de, 1632. *Copiosissima y verdadera relacion del incendio del monte Vesuvio, donde se da cuenta de veinte incendios quelha auido sin este ultimo*. Naples: Octavio Beltran.

Aziza, C. (ed.) 1995. *Pompéi, une anthologie composée et présentée par Claude Aziza*. Paris: Editions Pocket.

Balmuth, M. S., D. K. Chester, P. A. Johnston (eds) 2005. *Cultural responses to the volcanic landscape: the Mediterranean and beyond*. Colloquia and Conference Papers 8, Archaeological Institute of America, Boston.

Baratta, M. 1897. *Il Vesuvio e le sue eruzione*. Rome: Dante Alighieri.

—— 1901. *I terremoti d'Italia*. Turin: Fratelli Bocca.

Barberi, F., & M. L. Carapezza 1996. The Campi Flegrei case history. In *Monitoring and mitigation of volcano hazards*, R. Scarpa & R. I. Tilling (eds), 771–86, Berlin: Springer.

Barberi, F. & L. Leoni 1980. Metamorphic carbonate ejecta from Vesuvius Plinian eruptions: evidence of the occurrence of shallow magma chambers. *Bulletin Volcanologique* **43b**(1), 107–120.

Barberi, F., H. Bizouard, R. Clocchiatti, N. Metrich, R. Santacroce, A. Sbrana 1981. The Somma–Vesuvius magma chamber: a petrological and volcanological approach. *Bulletin Volcanologique* **44**(3), 295–315.

Barberi, F., G. Corrado, F. Innocenti, G. Luongo 1984. Brief chronicle of a volcano emergency in a densely populated area. *Bulletin Volcanologique* **47**, 175–85.

Barberi, F., D. Hill, F. Innocenti, G. Luongo, M. Treuil (eds) 1984. The bradyseismic crisis at Phlegraean Fields, Italy. *Bulletin Volcanologique* **47**(2; special issue), 173–412.

Barberi, F., C. Macedonio, M. T. Pareschi, R. Santacroce 1990. Mapping the tephra fallout risk: an example from Vesuvius (Italy). *Nature* **344**, 142–4.

Barberi, F., M. Rosi, R. Santacroce, M. F. Sheridan 1983. Volcanic hazard zonation: Mt Vesuvius. In *Forecasting volcanic events*, H. Tazieff & J. C. Sabroux (eds), 149–61. Amsterdam: Elsevier.

Baxter, P. J. 1990. Medical effects of volcanic eruptions, I: main cases of death and injury. *Bulletin of Volcanology* **52**, 532–44.

Beard, M. 2008. *Pompeii: the life of a Roman town*. London: Profile.

Beccaluva, L., G. Biancini, M. Wilson (eds) 2007. *Cenozoic volcanism in the Mediterranean area*. Special Paper 418, Geological Society of America, Boulder, Colorado.

Beneventano, Falcone, *c.* 1140. *Chronicon*. In Gasparini & Musella (1991: 175).

Berkeley, G. 1717. Extract of a letter of Mr Edw. Berkeley from Naples, giving several curious observations and remarks on the eruption of fire and smoke from Vesuvius. *Royal Society of London, Philosophical Transactions* **30**, 708–713.

Blunt, A. 1975. *Neapolitan Baroque and Rococo architecture*. London: Zwemmer.

Boccaccio, G. 1473. *De Montibus, silvis, fontibus, lacubis, fluminibus, stagnis seu paludibus et de nominibus maris*. Venice: Wendelin von Speyer.

Bonito, M. 1691. *Terra tremante ovvero continuatione de' terremoti dalla creatione del Mondo fino al tempo presente*. Naples: Forni [reissue 1980].

Bottoni, D. 1692. *Pyrologia topographica id est de igne dissertatio*. Naples.

Bourlot, J. 1867. *Etude sur le Vésuve*. Paris: Leiber.

Bowersock, G. W. 1978. The rediscovery of Herculaneum and Pompeii. *The American Scholar* **47**, 461–70.

Braccini, Giulio Cesare 1631. *Relazione dell'incendio fattosi nel Vesuvio alli 16 di decembre 1631, in una lettera all' Eminentissimo e Reverendissimo Signore il Signor Cardinal Girolamo Colonna*. Naples: Roncagliolo Secondino.

——— 1632. *Dell'incendio fattosi nel Vesuvio a XVI di dicembre MDCXXXI e delle cause e effetti con la narazione di quanto seguito in esso per tutto il marzo 1632 e con la storia di tutti gli altri incendi nel medismo monte avvenuti . . .* Naples: Roncaglio Secondino.

Breislak, S. & A. Winspeare 1794. *Memoria sul eruzione del Vesuvio accaduta la sera de 15 giunio 1794*, Naples.

Brigante, G., N. Spinosa, L. Stainton 1990. *In the shadow of Vesuvius: views of Naples from Baroque to Romanticism, 1631–1830.* [commentaries on the catalogue of the exhibition of the same title, Naples and London, 1990]. Naples: Electa.

Buchner, G. 1986. Eruzioni vulcaniche e fenomeni vulcano-tettonici di età preistorica e storica nell'isola d'Ischia. In Albore Livadie (1986b: 145–88).

Burke, E. 1757. *A philosophical enquiry into the origin of our ideas on the sublime and the beautiful*. London: Dodsley.

Butterworth, A. & R. Lawrence 2005. *Pompeii: the living city*. London: Weidenfeld & Nicolson.

Capasso, L. 2000. Herculaneum victims of the volcanic eruptions of Vesuvius in AD 79. *The Lancet* **356**, 1344–6.

Capece, Ascanio 1631–2. Lettere scritte al P. Antonio Capece della Compania di Gesù a Roma, 20 e 27 dicembre 1631 e 3 gennaio 1632. In Riccio (1889: 495–501).

Capocci, E., 1835. Nuove ricerche sul noto fenomeno delle colonne perforate dalle poladi nel tempio di Serapide in Pozzuoli. *Il Progresso delle Scienze Lettere delle Arte* **11**, 66–76.

Capuano, G. 1994. *Viaggiatori britannici a Naploi tra 500 e 600*. Salerno: Laveglia.

Carafa, Gregorio 1632. *In opusculum de novissima Vesuvii conflagratione: epistola isagogica*. Naples: Egidio Longo.

Carlino, S., R. Somma, G. C. Mayberry 2008. Volcanic risk perception of young people in the urban areas of Vesuvius: comparisons with other volcanic areas and implications for emergency management. *Journal of Volcanology and Geothermal Research* **172**, 229–43.

Carta, S., R. Figari, G. Sartoris, E. Sassi, R. Scandone 1981. A statistical model for Vesuvius and its volcanological implications. *Bulletin Volcanologique* **44**(2), 129–51.

Cassiodorus, M. A. [AD 537] 1894. *Cassiodori Senatoris variae* [notably letter 50] [Monumenta Germaniae Historica (Auctorum Antiquissimum) XII] Berlin: Weidmann.

Castaldo, A. 1538. [Extract from] *Diario delle cose accurse in Napoli al tempo del Vicerè Don Pedro da Toledo*. In Parascandola (1946: 185).

Ceraso, Francesco, 1632. *L' opre stupende e maravigliosi eccessi dalla natura prodotti nel Monte Vesuvio dell Citta di Napoli*. Naples: Roncagliolo Secondino.

Chateaubriand, F. R. de 1827. *Voyage en Italie*. Paris: Lavocat.

Chester, D. K. 1993. *Volcanoes and society*. London: Edward Arnold.

Chester, D. K. & A. M. Duncan 2007. Lieutenant-Colonel Delmé-Radcliffe's report on the 1906 eruption of Vesuvius, Italy. *Journal of Volcanology and Geothermal Research* **166**, 204–216.

Chester, D. K., C. J. L. Dibben, A. M. Duncan 2002. Volcanic hazard assessment in western Europe. *Journal of Volcanology and Geothermal Research* **115**, 411–35.

Chiesa, S., S. Poli, L. Vezzoli 1986. Studio dell'ultima eruzione storica dell'isola Ischia: la colata dell'Arso – 1302. *Bollettino Gruppo Nazionale di Vulcanologia*, 153–66.

Chiodini, G., R. Chioni, G. Magro, L. Marini, C. Panichi, B. Raco, M. Russo 1997. Chemical and isotopic variations of Bocca Grande fumarole (Solfatara volcano, Phlegraean Fields). *Acta Vulcanologica* **8**, 228–32.

Christian VIII of Denmark 1973. *Kong Christian VIII's Dagbøger og Optegnelser, bind II-I, 1815–1821, udg Af Det Kongelige Danske Selskab for Fædrelandets Historie* [King Christian VIII's diaries and notes, II-I, 1815–1821]. Copenhagen: Gyldendal.

Cioni, R., L. Civetta, P. Marianelli, N. Métrich, R. Santacroce, A. Sbrana 1995. Compositional layering and syn-eruptive mixing of a periodically refilled magma chamber: the AD 79 Plinian eruption of Vesuvius. *Journal of Petrology* **36**, 739–76.

Cioni, R., L. Gurioli, A. Sbrana, G. Vougioukalakis 2000. Precursors of the Plinian eruptions of Thera (Late Bronze Age) and Vesuvius (AD 79): data from archaeological areas. *Physics and Chemistry of the Earth* **25**, 719–24.

Cioni, R., R. Santacroce, A. Sbrana 1999. Pyroclastic deposits as a guide for reconstructing the multi-stage evolution of the Somma–Vesuvius caldera. *Bulletin of Volcanology* **60**, 207–222.

Civetta L. 1998. The emergency plans of the Neapolitan area. *Cities on Volcanoes, International Meeting.* G. Orsi, M. Di Vito, R. Isaia (eds), 83–5. Naples: Osservatorio Vesuviano.

Civetta L. & R. Santacroce 1992. Steady-state magma supply in the last 3400 years of Vesuvius activity. *Acta Vulcanologica* **2**, 147–60.

Civetta, L. L. Cuna, M. De Lucia, G. Orsi 2004. *Il Vesuvio negli occhi: storie di osservatori.* Osservatorio Vesuviano, Naples.

Civetta, L., G. Gallo, G. Orsi 1991. Sr and Nd isotope and trace-element constraints on the chemical evolution of the magmatic system of Ischia (Italy) in the last 55 ka. *Journal of Volcanology and Geothermal Research* **47**, 213–30.

Civetta, L., G. Orsi, L. Pappalardo, R. V. Fisher, G. Heiken, M. Ort 1997. Geochemical zoning, mingling, eruptive dynamics and depositional processes – the Campanian Ignimbrite, Campi Flegrei caldera, Italy. *Journal of Volcanology and Geothermal Research* **75**, 183–219.

Civetta, L. and 20 co-authors 1998. Volcanic, deformational and magmatic history of the Campi Flegrei caldera. In *Cities on Volcanoes, International Meeting,* G. Orsi, M. Di Vito, R Isaia (eds) 26–71. Naples: Osservatorio Vesuviano.

Cole, P. D. & C. Scarpati 1993. A facies interpretation of the eruption and emplacement mechanisms of the upper part of the Neapolitan Yellow Tuff, Campi Flegrei, southern Italy. *Bulletin of Volcanology* **55**, 311–26.

Cole, P. D., A. Perrotta, C. Scarpati 1994. The volcanic history of the southwestern part of the city of Naples. *Geological Magazine* **131**, 785–99.

Colucci Pescatori, G. 1986. Fonti antiche relative alla eruzioni vesuviane ed altri fenomeni vulcanici successivi al 79 DC. In Albore Livadie (1986b: 134–41).

———— 1986. Osservazioni su Abellinum tardo-antica e sull eruzione del 472 DC. In Albore Livadie (1986b: 121–33).

Constantine, D. 2001. *Fields of fire: a life of Sir William Hamilton.* London: Weidenfield & Nicholson.

Contreras, Alonso de 1990. *Mémoires du Capitán Alonso de Contreras* (translated by O. Aubertin). Paris: Viviane Hamy.

Cortini, M. & R. Scandone 1982. The feeding system of Vesuvius between 1754 and 1944. *Journal of Volcanology and Geothermal Research* **12**, 393–400.

Damien, St Peter, *c.* 1050. Opera Omnia: varia 19. In Gasparini & Musella (1991: 171).

D'Armes, J. 1970. *Romans in the Bay of Naples.* Cambridge, Mass.: Harvard University Press.

De Bottis, G. 1786. *Istoria di vari incendi del Monte Vesuvio.* Naples

Delibrias, G., G. M. Di Paola, M. Rosi, R. Santacroce 1979. La storia eruttiva del complesso vulcanico Somma Vesuvio ricostruita dalle successione piroclastiche del Montre Somma. *Rendiconti della Società Italiana di Mineralogia e Petrologia* **35**, 411–38.

Della Torre, G. M. 1755. *Storia e fenomini del Vesuvio.* Naples: Raimondi.

———— 1761. *Supplemento alla storia del Vesuvio ove si descrive l'incendio del 1760.* Naples: Raimondi.

———— 1768. *Storia e fenomeni del Vesuvio (con supplemento).* Naples: Raimondi.

———— 1777. *Scienza della natura generale (II).* Naples: Raimondi.

Delli Falconi, Marco Antonio, 1539. *Dell'incendio di Pozzuoli nel MDXXXVIII.* Naples: Sultzbach.

De Lorenzo 1906. The eruption of Vesuvius in April 1906. *Geological Society of London, Quarterly Journal* **62**, 476–83.

Del Nero, Francesco 1846. Sul terremoto di Pozzuoli dal quale ebbe origine la montagna nuova, nel 1538. In *Documenti relativi al tempo e al governo di Don Pedro da Toledo, Vice-Re di Napoli dal 1532 al 1553.* Archivio Storico Italiano 9.

Denon, Dominique-Vivant 1997. *Voyage au royaume de Naples.* Paris: Perrin.

De Seta, C., L. Di Mauro, M. Perone 1980. *Ville vesuviane*. Milan: Rusconi.

De Vita, S., and 15 other authors 1999. The Agnano–Monte Spina eruption (4100 years BP) in the restless Campi Flegrei caldera. *Journal of Volcanology and Geothermal Research* **91**, 269–301.

De Vivo, B., R. Scandone, R. Trigila (eds) 1992. *Mount Vesuvius. Journal of Volcanology and Geothermal Research* **58**(special issue).

Dickens, C. 1846. *Pictures from Italy* [serialized in the *Daily News*, London, 1846]. Harmondsworth: Penguin [reissue 1998].

Dipartimento della Protezione Civile, Regione Campania 1984. *Elementi di base per la pianificazione nazionale di emergenza dell'area zone Flegrea*. Prefettura di Napoli, Naples.

—— 1995 (and 2005). *Pianificazione Nazionale d'Emergenza dell'Area Vesuviana*. Prefettura di Napoli, Naples.

—— 2006. *Mesimex 2006*. Prefettura di Napoli, Naples.

Dio Cassius [AD 207–219] 1961. *Dio's Roman history* (translated by E. Cary) [notably 66.21–6, 67, 77.2.1]. Cambridge, Mass.: Harvard University Press.

Diodorus Siculus. *c.* 30 BC 1960. *Library of history* (12 vols; translated by C. H. Oldfather) [notably 4.21.5]. Cambridge, Mass.: Harvard University Press.

Di Vito, M. A., R Isaia, G. Orsi, J. Southon, S. de Vita, M. D'Antonio, L. Pappalardo, M. Piochi 1999. Volcanism and deformation since 12000 years at the Campi Flegrei caldera (Italy). *Journal of Volcanology and Geothermal Research* **91**, 221–46.

Di Vito, M., L. Lirer, G. Mastrolorenzo, G. Rolandi 1987. The 1538 Monte Nuovo eruption (Campi Flegrei, Italy). *Bulletin of Volcanology* **49**, 608–15.

Dobran, F. 1993. *Global volcanic simulation of Vesuvius*. Pisa: Giardini.

Dobran, F. 1994. Incontro con il Vesuvio. *Sapere* (November 1994), 11–16.

—— 1998. *Educazione al rischio Vesuvio*. Global Volcanic and Environmental Systems Simulation (GVES), Naples.

—— (ed.) 2006. *Vesuvius: education, security and prosperity*. Amsterdam: Elsevier.

Dobran, F. & G. Luongo 1995. *Vesuvius 2000*. Global Volcanic and Environmental Systems Simulation (GVES), Rome.

Dobran, F., F. Neri, M. Tedesco 1994. Pyroclastic flow hazard at Vesuvius. *Nature* **367**, 551–4.

Dumas, A. 1984. *Le corricolo*. Paris: Desjonquères. [First published in 1843 as *Corricolo: impressions de voyage en Italie*. Paris: Dolin.]

Dupaty, C. 1826. *Lettres sur l'Italie*. Paris: Veldière.

Dvorak, J. J. & G. Berrino 1991. Recent ground movement and seismic activity in Campi Flegrei, southern Italy: episodic growth of a resurgent dome. *Journal of Geophysical Research* **96**, 2309–323.

Dvorak, J. J. & P. Gasparini 1991. History of earthquakes and vertical ground movement in Campi Flegrei caldera, southern Italy: comparison of precursory events to the AD 1538 eruption of Monte Nuovo and of activity since 1968. *Journal of Volcanology and Geothermal Research* **48**, 77–92.

Estrada, Diego, Duque de 1982. *Comentarios del desengañado de si mismo: vida del mismo autor, Diego Duque de Estrada. (Edición, introducción, y notas de I. Henry Ettinghausen)* Madrid: Castalia.

Evelyn, John 1819. *Diary* (2 vols; William Bray, ed.) London: Everyman.

Falcone, S. 1632. *Discorso naturale delle cause et effetti causati nell'incendio del Monte Vesuvio*. Napoli: Ottavio Beltrano.

Fisher, R. V., G. Orsi, M. Ort, G. Heiken 1993. Mobility of a large-volume pyroclastic flow: emplacement of the Campanian ignimbrite. *Journal of Volcanology and Geothermal Research* **59**, 205–220.

Florus, L. A. [AD 116] 1984. *Epitome of Roman history* (translated by E. S. Forster) [notably 2.8]. Cambridge, Mass.: Harvard University Press.

Fothergill, B. 1969. *Sir William Hamilton*. London: Faber & Faber.

—— 2005. *Sir William Hamilton: envoy extraordinary*. Stroud: Tempus.

Franchi dall'Orto, L. 1993. *Ercolano 1738–1988: 250 anni di ricerca archeologica*. Roma: Bardi.

Franco, D. 1872. Sur l'éruption d'avril 1872 au Vésuve. *Académie des sciences de Paris, Comptes Rendus* **75**(2), 221–4.

Frat'Angelo (de Eugenij da Perugia) 1631. *Il maraviglioso e tremendo incendio del Monte Vesuvio; detto à Napoli la Montagna Somma nel 1631*. Naples: Ottavio Beltrano.

Frontinus [AD 84–96] 1961. *Strategemata* (translated by C. E. Bennett) [notably 1.5.21]. Cambridge, Mass.: Harvard University Press.

Fucini, R. 1878. *Napoli ad occhio nudo*. Florence: Le Monnier [reissued 2004 – Naples: Avagliano].

Gallet, Y., A. Genevey, M. Le Goff 2002. Three millennia of directional variation of the Earth's magnetic field in western Europe as revealed by archeological artefacts. *Physics of the Earth and Planetary Letters* **131**, 81–9.

Gasparini, P. & S. Musella 1991. *Un viaggio al Vesuvio*. Naples: Liguori.

Giacomelli, L. & R. Scandone 1992. *Campi Flegrei, Campania felix, I: il Golfo di Napoli tra storia ed eruzioni*. Naples: Liguori.

———— 1992. *Campi Flegrei, Campania felix, II: guida alle escursioni dei vulcani napoletani*. Naples: Liguori.

Giacomelli, L., A. Perrotta, R. Scandone, C. Scarpati 2003. The eruption of Vesuvius of AD 79 and its impact on human environment in Pompei [*sic*]. *Episodes* **26**, 234–7.

Gialanella, P. R., A. Incoronato, F. Russo, P. Sarno, G. Nigro 1993. Magnetic stratigraphy of Vesuvius products, I: 1631 lavas. *Journal of Volcanology and Geothermal Research* **58**, 203–209.

Gialanella, P. R., A. Incoronato, F. Russo, P. Sarno, A. D. Martino 1998. Magnetic stratigraphy of Vesuvius products, II: Medieval lavas. *Quaternary International* **47/48**, 135–8.

Gigante, M. 1989. *Il fungo sul Vesuvio secondo Plinio il Giovane*. Naples: Lucarini.

Giubelli, G. [undated]. *Oplontis: la Villa di Poppaea*. Naples: Carcavallo.

Giuliani, Gianbernadino 1632. *Trattato del Monte Vesuvio e de' suoi incendi*. Naples: Egidio Longo.

Gleijeses, V. 1980. *Ville e palazzi vesuviani*. Naples: Società Editrice Napoletana.

Goethe, J. W. von 1810–1817. *Aus meinem Leben: Dichtung und Wahrheit*, Teil 4 [Goethe sämtliche Schriften 1810–1817, Band 26]. Vienna: Anton Strauß. [*From my life: poetry and truth*, part 4 [Goethe's complete scripts 1810–1817, vol. 26] [Published as *Italian journey* (translated by W. H. Auden & E. Meyer). Harmondsworth: Penguin, 1970].

Guest, J. E., P. D. Cole, A. M. Duncan, D. K. Chester 2003. *The volcanoes of southern Italy*. London: Geological Society of London.

Guzzo, P. G. (ed.) 2003. *Storia da un'eruzione: Pompei, Ercolano, Oplontis*. Naples: Electa.

Hamilton, W. 1767. Two letters . . . containing an account of the last eruption of Mount Vesuvius. *Royal Society of London, Philosophical Transactions* **57**, 192–200.

———— 1768. An account of the eruption of Mount Vesuvius in 1767. *Royal Society of London, Philosophical Transactions* **58**, 1–14.

———— 1769. A letter . . . containing some farther [*sic*] particulars on Mount Vesuvius and other volcanos in the neighbourhood. *Royal Society of London, Philosophical Transactions* **59**, 18–22.

———— 1770. An account of a journey to Mount Etna. *Royal Society of London, Philosophical Transactions* **60**, 1–19.

———— 1771. Remarks upon the nature of the soil of Naples and its neighbourhood. *Royal Society of London, Philosophical Transactions* **61**, 1–48.

Hamilton, Sir W. 1774. *Observations on Mount Vesuvius, Mount Etna and other volcanos of the Two Sicilies* [a series of letters addressed to the Royal Society of London]. London: T. Cadell.

———— 1776/1779. *Campi Phlegraei: observations on the volcanos of the Two Sicilies* [with a supplement published in 1779]. Naples: Fabris.

———— 1780. An account of an eruption of Mount Vesuvius which happened in August 1779. *Royal Society of London, Philosophical Transactions* **70**, 42–84.

———— 1783. An account of the earthquake which happened in Italy from February to May 1783. *Royal Society of London, Philosophical Transactions* **73**, 169–208.

———— 1786. The present state of Mount Vesuvius; with an account of a journey into the province of Abruzzo, and a voyage to the island of Ponza. *Royal Society of London, Philosophical Transactions* **76**, 365–81.

———— 1795. An account of the late eruption of Mount Vesuvius. *Royal Society of London, Philosophical*

Transactions **85**, 73–116.

Heather, P. 2005. *The fall of the Roman Empire: a new history*. London: Macmillan.

Hine, H. M. 2002. Seismology and vulcanology in antiquity. In *Science and mathematics in ancient Greek culture*, C. J. Tuplin & T. E. Rihll (eds), 56–75. Oxford: Oxford University Press.

Homer *c.* 750 BC. *Odyssey* (translated by E. V. Rieu 1946) [notably Book 11]. Harmondsworth: Penguin.

Howatson, M. C. & I. Chivers (eds) 1993. *Concise Oxford companion to classical literature*. Oxford: Oxford University Press.

Imbò, G. 1949. L'attività eruttiva vesuviana e relative osservazioni nel corso dell'intervallo intereruttivo 1906–1944 ed in particolare del parossismo del marzo 1944. *Annali dell'Osservatorio Vesuviano* **5**, 185–380.

———— 1965. *Catalogue of the active volcanoes of the world including Solfatara Fields, Part XVII: Italy*. Rome: International Association of Volcanology.

Ingamells, J. 1997. *A dictionary of British and Irish travellers in Italy 1701–1800*. New Haven: Yale University Press.

Jashemski, W. F. 1979. Pompeii and Mount Vesuvius. In *Volcanic activity and human ecology*, P. D. Sheets & D. K. Grayson (eds), 587–622. New York: Academic Press.

———— 2002. The Vesuvian sites before AD 79: the archaeological literary and epigraphical evidence. In Jashemski & Meyer (2002: 6–28).

Jashemski, W. F. & F. G. Meyer (eds) 2002. *The natural history of Pompeii: a systematic survey*. Cambridge: Cambridge University Press.

Jenkins, I. & K. Sloan 1996. *Vases and volcanoes: Sir William Hamilton and his collection*. London: British Museum Press.

Johnston-Lavis, H. J. 1884. The geology of Monte Somma and Vesuvius, being a study in volcanology. *Geological Society of London, Quarterly Journal* **40**, 35–149.

———— 1891. *The southern Italian volcanoes*. Naples: Furchheim.

———— 1909. The eruption of Vesuvius in April 1906. *Royal Dublin Society, Transactions* **9**, 139–200.

Johnston-Lavis, H. J. & A. Lavis 1918. *Bibliography of the volcanoes of southern Italy*. London: University of London Press.

Jones, F. L. 1964. *The letters of Percy Bysshe Shelley* (2 vols). Oxford: Oxford University Press.

Keller, J., W. B. F. Ryan, D. Ninkovitch, R. Altherr 1978. Explosive volcanic activity in the Mediterranean over the past 200 000 years as recorded in deep-sea sediments. *Geological Society of America, Bulletin* **89**, 591–604.

Kilburn, C. J. & W. McGuire 2001. *Italian volcanoes*. Harpenden: Terra.

Lacroix, A. 1906. *L'éruption du Vésuve en avril 1906*. Paris: Armand Colin.

———— 1908. *La Montagne Pelée après ses éruptions, avec observations sur les éruptions du Vésuve en 79 et en 1906*. Paris: Masson.

Lamartine, A. de 1849. *Le dernier chant du pèlerinage d'Harold* (section XII). Paris: Alphonse Lemerre.

Le Hon, H. S. 1865. *Histoire complète de la grande éruption du Vésuve de 1631*. Brussels: M. Hayez. [See also: *Académie Royale de Belgique, Bulletin* Série 2, **20**(8), 1–64.].

Leone, A. 1514 (1934). *De Nola* [notably 1.1]. Venice: By permission of the Doge. In Gasparini & Musella (1991: 173).

Leone, U. 2006. *Sicurezza ambientale: La Campania – gli assi portanti* (number 8). Naples: A. Guida.

Lewis, N. 1978. *Naples '44*. London: Eland.

Lirer, L., R. Munno, I. Postiglione, A. Vinci, L. Vitelli 1997. The AD 79 eruption as a future explosive scenario in the Vesuvian area: evaluation of associated risk. *Bulletin of Volcanology* **59**, 112–24.

Lirer, L., T. Pescatore, B. Booth, G. P. L. Walker 1973. Two Plinian pumice-fall deposits from Somma–Vesuvius, Italy. *Geological Society of America, Bulletin* **84**, 759–72.

Lirer, L., P. Petrosino, I. Alberico 2001. Hazard assessment at volcanic fields: the Campi Flegrei case history. *Journal of Volcanology and Geothermal Research* **112**, 53–73.

Lirer, L., P. Petrosino, I. Alberico, I. Postiglione 2001. Long-term volcanic hazard forecasts based on Somma–Vesuvius past eruptive activity. *Bulletin of Volcanology* **63**, 45–60.

Lobley, J. L. 1889. *Mount Vesuvius*. London: Roper & Drowley.

Luongo, G. & A. Mazzarella 2002. On the time-scale variations of the eruptive activity of Vesuvius. *Journal of Volcanology and Geothermal Research* **113**, 1–3.

Luongo, G. & R. Scandone (eds) 1991. Campi Flegrei. *Journal of Volcanology and Geothermal Research* (special issue, whole volume) **48**.

Luongo, G., E. Cubellis, F. Obrizzo, S. Petrazzuoli 1991. A physical model for the origin of volcanism of the Tyrrhenian margin: the case of the Neapolitan area. *Journal of Volcanology and Geothermal Research* **48**, 173–86.

Luongo, G., A. Perrotta, C. Scarpati 2003. Impact of the AD 79 explosive eruption on Pompeii, I: relations among the depositional mechanisms of the pyroclastic products, the framework of the buildings, and the associated destructive events. *Journal of Volcanology and Geothermal Research* **125**, 203–223.

Luongo, G., A Perrotta, C. Scarpati, E. De Carolis, G. Patricelli, A. Ciarello 2003. Impact of the AD 79 explosive eruption on Pompeii, II: causes of death of the inhabitants inferred by stratigraphic analysis and aerial distribution of the human casualties. *Journal of Volcanology and Geothermal Research* **125**, 169–200.

Lyell, Sir Charles 1830. *Principles of geology*. London: Murray.

Macedonio, G., M. T. Pareschi, R. Santacroce 1990, Renewal of explosive activity at Vesuvius: models for expected fallout. *Journal of Volcanology and Geothermal Research* **40**, 327–42.

Manso, Giovanni Battista, 1632. Lettere del Sig. Giovanni Battista, Marchese di Villa, scritte da Napoli al Sig. Antonio Bruni in materia del Vesuvio. In Riccio (1889: 502–18).

Marcellinus, Comes [Count] [*c.* 530] 1893/94. *Chronicon* [Monumenta Germaniae Historica (Auctorum Antiquissimum) XI: 90].

Marchesino, Francesco 1538. Copia di una lettera di Napoli che contiene li stupendi et gran prodigi apparsi à Pozzolo. In Parascandola (1946: 176–7).

Marcus Aurelius (AD 166–178) 1916. *Ad se ipsum* (translated by C. R. Haines). Cambridge, Mass.: Harvard University Press.

Martial (AD 88) 1925. *Epigrams* (2 vols, translated by W. A. Ker) [notably 4.44]. Cambridge, Mass.: Harvard University Press.

Massood, E. 1995. Row erupts over evacuation plans for Mount Vesuvius. *Nature* **377**, 471.

Mastrolorenzo G., R. Munno, G. Rolandi 1993. Vesuvius 1906: a case study of a paroxysmal eruption and its relation to eruptive cycles. *Journal of Volcanology and Geothermal Research* **58**, 217–37.

Mastrolorenzo, G., D. M. Palladino, G. Vecchio, J. Taddeucci 2002. The AD 472 Pollena eruption of Somma–Vesuvius and its environmental impact at the end of the Roman Empire. *Journal of Volcanology and Geothermal Research* **113**, 19–36.

Mastrolorenzo, G., P. Petrone, P. Pagano, A. Incoronato, P. J. Baxter, A. Canzanella, L. Fattore 2001. Herculaneum victims of Vesuvius in AD 79. *Nature* **410**, 769–70.

Mastrolorenzo, G., P. Petrone, L. Pappalardo, M. F. Sheridan 2006. The Avellino 3780-yr-BP catastrophe as a worst-case scenario for a future eruption at Vesuvius. *National Academy of Sciences, Proceedings* **103**, 4366–70.

McKenzie, D. P. 1970. Plate tectonics of the Mediterranean region. *Nature* **226**, 239–43.

Mecatti, G. M. 1761. *Continuazione delle osservazioni sopra diverse eruzione del Vesuvio* Naples: De Simoni.

Mercalli, G. 1891. I terremoti napolitani del secolo XVI ed un manuscritto inedito di Cola Aniello Pacca. *Bollettino della Società Geologica Italiana* **10**, 179–96.

——— 1906. L'eruzione vesuviana dell'aprile 1906. *Natura ed Arte*, *1906*, 763–72.

——— 1907. *I vulcani attivi della terra*. Milan: Ulrico Hoepli.

Mercalli, G., M. Baratta, B. Friedlander, A. Aguilar, O. Scarpa 1907. *Il Vesuvio e la grande eruzione dell'aprile 1906*. Napoli: Colavecchia, Colombai.

Métrich, N. 1985. *Mécanismes d'évolution à l'origine des roches volcaniques potassiques d'Italie centrale et méridionale: exemples du Mt Somma–Vésuve, des Champs Phlégréens et de l'île de Ventotene*. Thèse, Université de Paris-Sud.

Miccio, Scipione 1600. *Vita di Don Pietro da Toledo* [Archivio Storico Italiano 9]. Rome [reissue 1846].

Monticelli, T. & N. Covelli 1823. *Storia de' fenomeni dei Vesuvio*. Naples.

Mormile, G. 1632. *L'incendio del Monte Vesuvio*. Napoli: Egidio Longo.

Nazzaro, A. 2001. *Il Vesuvio: storia eruttiva e teorie vulcanologiche* (2nd edn). Naples: Liguori.

———— 1998. Some considerations on the state of Vesuvio in the Middle Ages and the precursors of the 1631 eruption. *Annali Geofisica* **41**, 555–65.

———— 1999. The Vesuvius shape before 79 AD according to a new finding from a Pompei fresco and Vesuvio central cone history in the last 200 years. *Annali Geofisica* **42**, 715–23.

Newhall, C. G. & S. Self 1982. The volcanic explosivity index (VEI): an estimate of explosive magnitude for historical volcanism. *Journal of Geophysical Research (Oceans and Atmosphere)* **87**, 1231–8.

Orsi, G., L. Civetta, C. Del Gaudio, S. De Vita, M. A. Di Vito, R. Isaia, S. M. Petraluzzuoli, G. P. Ricciardi, C. Ricco 1999. Short-term ground deformations and seismicity in the resurgent Campi Flegrei caldera (Italy): an example of active block resurgence in a densely populated area. *Journal of Volcanology and Geothermal Research* **91**, 415–51.

Orsi, G., S. De Vita, M. Piochi 1998. The volcanic island of Ischia. In *Field excursion guide: Cities on volcanoes* [international meeting], G. Orsi, M. Di Vito, R. Isaia (eds), 72–8. Naples: Osservatorio Vesuviano.

Orsi, G., M. Piochi, L. Campajola, A. D'Onofrio, L. Gianella, F. Terrasi 1996. ^{14}C geochronological constraints for the volcanic history of the island of Ischia (Italy) over the last 5000 years. *Journal of Volcanology and Geothermal Research* **71**, 249–57.

Pacca, C. A. *c.* 1580. *Discorso di terremoti*. In Mercalli (1891).

Palmieri, L. 1872. *Incendio Vesuviano del 26 aprile 1872*. Torino: Fratelli Bocca.

———— 1873. *The eruption of Vesuvius in 1872* (translated by R. Mallet). London: Asher.

———— 1887. Il Vesuvio e la sua storia. In *Lo spettatore del Vesuvio e dei Campi Flegrei*, Club Alpino Italiana (ed.). Naples: Presso Federico Furchheim.

Palumbo, A. 1999. The activity of Vesuvius in the next millennium. *Journal of Volcanology and Geothermal Research* **88**, 125–9.

———— 2003. *Il Vesuvio, I Campi Flegrei, e I Napolitani*. Naples: Liguori.

Parascandola, A. 1946. Il Monte Nuovo ed il Lago Lucrino. *Bollettino della Società dei Naturalisti in Napoli* **55**, 152–264.

Pareschi, M. T., L. Cavarra, M. Favalli, F. Giannini, A. Meriggi 2000. GIS and volcanic risk management. *Journal of Natural Hazards* **21**, 361–79.

Paterne, M., F. Guichard, J. Labeyrie 1988. Explosive activity of the south Italian volcanoes during the past 80000 years as determined by marine tephrochronology. *Journal of Volcanology and Geothermal Research* **34**, 153–72.

Paul the Deacon [*c.* 780] 1907. *Historia Langobardorum* ["history of the Lombards"] (translated by W. D. Foulke) [notably 6.9]. Philadelphia: University of Pennsylvania Press.

Perret, F. A. 1924. *The Vesuvius eruption of 1906: study of a volcanic cycle*. Publication 339, Carnegie Institution, Washington DC.

Phillips, J. 1869. *Vesuvius*. Oxford: Oxford University Press.

Pighius, S. 1587. *Hercules prodicius seu principiis iuventutis vita et peregrinatio*. Antwerp: C. Plautinus.

Pindar 1997. *Pythian odes* (translated by W. H. Race) [notably 1.5.17–19]. Cambridge, Mass.: Harvard University Press.

Pliny the Elder [Gaius Plinius Secundus, *c.* AD 77] 1940–1963. *Natural History* 2.199; 2.203. (translated by H. Rackham, W. H. S. Jones, L. F. Newman). Cambridge, Mass.: Harvard University Press.

Pliny the Younger [Gaius Plinius Caecilius Secundus, *c.* AD 106–107] 1969. *Letters and Panegyricus* 1: book VI, letters 16 and 20 (translated by B. Radice). Cambridge, Mass.: Harvard University Press.

Plutarch [*c.* AD 90] 1916. *The parallel lives: Crassus* [9.1–3] (11 vols, translated by B. Perrin). Cambridge, Mass.: Harvard University Press.

Porzio, Simone 1538. *De conflagratione agri puteoliani Simoni Portii Epistula*. Naples: G. Sultzbach.

Preusse, P. 1934. Ein Wort zur Vesugvestalt und Vesuvtätigkeit im Altertum. *Klio* **27**, 295–310.

———— 1936. Considérations sur la forme du Vésuve et son activité dans l'antiquité. *Bulletin Volcan-*

ologique **VIII**, 11–23.

Principe, C. 1998. The 1631 eruption of Vesuvius: volcanological concepts in Italy at the beginning of the XVIIth century. In *Volcanoes and history*, Proceedings of the 20th International Commission on the History of the Geological Sciences Symposium Napoli–Eolie–Catania (Italy), N. Morello (ed.), 525–42. Genoa: Brigati.

Principe, C., A. Paiotti, J. C. Tanguy, M. Le Goff 1998. Archéomagnétisme et chronologie des éruptions du Vésuve. *Société Géologique de France, Bulletin de la Section de Volcanologie* **44**, 25.

Principe, C., J. C. Tanguy, S. Arrighi, A. Paiotti, M. Le Goff, U. Zoppi 2004. Chronology of Vesuvius activity from AD 79 to 1631 based on archeomagnetism of lavas and historical sources. *Bulletin of Volcanology* **66**, 703–724.

Procopius [*c.* AD 536–553] 1919. *History of the wars: the Gothic war* (7 vols; translated by H. B. Dewing) [notably Books 6 and 8]. Cambridge, Mass.: Harvard University Press.

Recupito, G. C. 1632. *De Vesuviano incendio nuntius*. Naples: Egidio Longo.
——— 1635. *Avviso dell'incendio del Vesuvio*. Naples: Egidio Longo.

Renna, E. 1992. *Vesuvius Mons: aspetti del Vesuvio nel mondo antico, tra filologica, archeologia, vulcanologia*. Naples: Procaccini.

Riccio, L. 1889. *Nuovi documenti sull'incendio vesuviano dell'anno 1631 e bibliografia di quella eruzione* [Archivio Storico per le Province Napolitane 14]. Naples: Giannini e Figli.

Rittmann, A. 1930. Geologie der Insel Ischia. *Zeitschrift für Vulkanologie* **6**, 1–265.
——— 1962. *Volcanoes and their activity*. New York: John Wiley.

Rittmann, A. & V. Gottini 1980. L'Isola d'Ischia. *Bollettino della Società Geologica Italiana* **101**, 131–274.

Rolandi, G. & F. Russo 1993. L'eruzione del Vesuvio del 1631. *Società Geologica Italiana, Bollettino* **112**, 315–32.

Rolandi, G., A. M. Barrella, A. Borrelli 1993. The 1631 eruption of Vesuvius. *Journal of Volcanology and Geothermal Research* **58**, 183–201.

Rolandi, G., P. Petrosini, J. McGeehin 1998. Interplinian activity at Somma–Vesuvius in the last 3500 years. *Journal of Volcanology and Geothermal Research* **82**, 19–52.

Rosi, M. & R. Santacroce 1983. The AD 472 "Pollena" eruption: volcanological and petrological data for this poorly known Plinian-type event at Vesuvius. *Journal of Volcanology and Geothermal Research* **17**, 249–71.

——— 1984. Volcanic hazard assessment in the Phlegraean Fields: a contribution based on stratigraphic and historical data. *Bulletin of Volcanology* **47**, 359–70.

——— 1986. L'attività del Somma–Vesuvio precedente l'eruzione del 1631: dati stratigrafici e vulcanologici. In Albore Livadie (1986b: 15–33).

Rosi, M. & A. Sbrana (eds) 1987. Phlegraean Fields. *Quaderni de La Ricerca Scientifica* **114**, 1–175.

Rosi, M., C. Principe, R. Vecci 1993. The 1631 Vesuvius eruption: a reconstruction based on historical and stratigraphical data. *Journal of Volcanology and Geothermal Research* **58**, 151–82.

Rosi, M., A. Sbrana, C. Principe 1983. The Phlegraean Fields: structural evolution volcanic history and eruptive mechanisms. *Journal of Volcanology and Geothermal Research* **17**, 273–88.

Rosi, M., L. Vezzoli, P. Aleotti, M. De Censi 1996. Interaction between caldera collapse and eruptive dynamics during the Campanian ignimbrite eruption, Phlegraean Fields, Italy. *Bulletin of Volcanology* **57**, 541–54.

Russo, A. 1818. Account. In Panvini, P. *Il forestiero alle antichità e curiosità naturali de Pozzuoli, Cuma, Baia, e Miseno*. Naples.

Saint-Non, Abbé de (Jean-Claude Richard) 1781–1787. *Voyage pittoresque de Naples et de Sicile* (4 volumes). Paris.

Sallust [*c.* 39 BC] 1921. *Historiae* (translated by J. C. Rolfe). Cambridge, Mass.: Harvard University Press.

Sampaolo, V. 1986. Dati archeologici e fenomeni vulcanici nel area Nolana. Nota preliminare. In Albore Livadie (1986b: 113–19).

Santacroce, R. 1983. A general model for the behaviour of the Somma–Vesuvius volcanic complex. *Journal of Volcanology and Geothermal Research* **17**, 237–48.

———— (ed.) 1987. Somma–Vesuvius. *Quaderni della Ricerca Scientifica* **114**(8) [plus geological map].

———— 1996. Preparing Naples for Vesuvius. *International Association of Volcanology and Chemistry of the Earth's Interior (IAVCEI), News* **1–2**, 5–7.

Scandone, R., F. Bellucci, L. Lirer, G. Rolandi 1991. The structure of the Campanian plain and the activity of the Neapolitan volcanoes. *Journal of Volcanology and Geothermal Research* **48**, 1–31.

Scandone, R., G. Arganese, F. Galdi 1993a. The evaluation of volcanic risk in the Vesuvian area. *Journal of Volcanology and Geothermal Research* **58**, 263–71.

Scandone, R., L. Giacomelli, P. Gasparini 1993b. Mount Vesuvius: 2000 years of volcanological observations. *Journal of Volcanology and Geothermal Research* **58**, 5–25.

Scarpati, C., P. Cole, A. Perrotta 1993. The Neapolitan yellow tuff: a large volume multiphase eruption from Campi Flegrei, southern Italy. *Bulletin of Volcanology* **55**, 343–56.

Scarth A. 1994. *Volcanoes*. London: UCL Press.

———— 1999. *Vulcan's fury: Man against the volcano*. London: Yale University Press.

———— 2000. The volcanic inspiration of some images in the *Aeneid*. *Classical World* **93**(6), 591–605.

———— 2002. A perspective on volcanic catastrophes. *European Geologist* **14**, 36–9.

———— 2002. *La catastrophe: Mount Pelée and the destruction of Saint-Pierre, Martinique*. Harpenden: Terra.

———— 2007. Dealing with Vesuvius. *European Geologist* **29**, 5–9.

———— 2007. Att Handskas med vulkanen Vesuvius. *Geologiskt Forum* **55**, 6–7.

Scarth, A. & J. C. Tanguy 2001. *Volcanoes of Europe*. Harpenden: Terra.

Scrope, G. J. Poulett 1823. An account of the eruption of Vesuvius in October 1822. *Quarterly Journal of Science (London)* **15**, 175–83.

Scrope, G. J. Poulett 1825. *Considerations on volcanos, the probable causes of their phenomena and their connection with the present state and past history of the globe; leading to the establishment of a new theory of the Earth*. London: W. Phillips.

Seneca, [AD 62–63] 1971. *Natural questions* (translated by T. H. Corcoran) [notably 6.1.1–3, 6.12.2, 6.27–1, 6.28.1]. Cambridge, Mass.: Harvard University Press.

Serao, F. 1738. *Istoria dell'incendio del Vesuvio accaduto nel mese di maggio dell'anno 1737, scritta per l'Accademia delle Scienze*. Naples: De Bonis.

———— 1743. *The natural history of Mount Vesuvius with the explanation of the various phenomena that usually attend the eruptions of the celebrated volcano*. London: E. Cave.

Shelley, P. B. 1912. *The letters* (2 vols; F. L. Jones [ed.]). Oxford: Oxford University Press.

Sheridan, M. F. & M. C. Malin 1983. Application of computer-assisted mapping to volcanic hazard evaluation of surge eruptions: Vulcano, Lipari and Vesuvius. *Journal of Volcanology and Geothermal Research* **17**, 187–202.

Sheridan, M. F., F. Barberi, M. Rosi, R. Santacroce 1981. A model for Plinian eruptions of Vesuvius. *Nature* **289**, 282–5.

Sigurdsson, H. S. 1999. *Melting the Earth: the history of ideas on volcanic eruptions*. New York: Oxford University Press.

———— 2002. Mount Vesuvius before the disaster. In *The natural history of Pompeii: a systematic survey*, W. F. Jashemski & F. G. Meyer (eds), 29–36. Cambridge: Cambridge University Press.

Sigurdsson, H. S. & S. Carey 2002. The eruption of Vesuvius in AD 79. In Jashemski & Meyer (2002: 37–64).

Sigurdsson, H. S., S. Carey, W. Cornell, T. Pescatore 1985. The eruption of Vesuvius in AD 79. *National Geographic Research* **1**, 332–87.

Sigurdsson, H. S., S. Cashdollar, R. S. J. Sparks 1982. The eruption of Vesuvius in AD 79: reconstruction from historical and volcanological evidence. *American Journal of Archaeology* **86**, 39–51.

Simkin, T. & L. Siebert 1994. *Volcanoes of the world* (2nd edn). Smithsonian Institution, Global Volcanism Program. Tucson, Arizona: Geoscience Press.

Sontag, S. 1992. *The volcano lover*. London: Jonathan Cape.

Sorrentino, I. 1734. *Istoria del Monte Vesuvio*. Naples: G. Severini.

Spallanzani, L. 1792. *Viaggi alle Due Sicilie e in alcune parti del' Appennino* (6 vols). Pavia: Stamperia Badassare Comini.

Spera, F. J., B. De Vivo, R. A. Ayuso, H. E. Belkin (eds) 1998. *Vesuvius. Journal of Volcanology and*

Geothermal Research **82** special issue [whole volume].

Statius [AD 93 to AD 95] 2003. *Silvae* (translated by D. R. Shackleton-Bailey). Cambridge, Mass.: Harvard University Press.

Stothers, R. B. & M. R. Rampino 1983. Volcanic eruptions in the Mediterranean before AD 630 from written and archaeological sources. *Journal of Geophysical Research* **88**, 6357–71.

Strabo [*c.* AD 18] 1923 *The geography* (8 vols, translated by H. L. Jones) [notably 1.2.18, 5.4.6, 5.4.8, 5.4.9]. Cambridge, Mass.: Harvard University Press.

Suetonius [AD 119 to 122] 1989. *Lives of the twelve Caesars* (translated by Robert Graves). Harmondsworth: Penguin.

Swinbourne, H. 1783. *Travels in the two Sicilies in the years 1777, 1778, 1779 and 1780.* London: Emsley.

Tacitus [AD 106 to AD 117] 1989. *Annals of imperial Rome* 1989 (translated by M. Grant 1989] [notably 4.67; 15.22; 15.34]. Harmondsworth: Penguin.

Tanguy, J. C. 1997. Le Vésuve et Naples: un cas critique de prévention du risque volcanique. *Société Géologique de France, Bulletin de la Section de Volcanologie* **41**, 1–9.

Tanguy, J. C., M. Le Goff, V. Chillemi, A. Paiotti, C. Principe, S. La Delfa, G. Patanè 1999. Variation séculaire de la direction du champ géomagnétique enregistrée par les laves de l'Etna et du Vésuve pendant les deux derniers millénaires. *Académie des Sciences de Paris, Comptes Rendus* **329**, 557–64.

Tanguy, J. C., M. Le Goff, C. Principe, S. Arrighi, V. Chillemi, A. Paiotti, S. La Delfa, G. Patanè 2003. Archeomagnetic dating of Mediterranean volcanics of the last 2100 years: validity and limits. *Earth and Planetary Science Letters* **211**, 111–24.

Toleto, Pietro Giacomo 1539. *Ragionamento del terremoto, del Nuovo Monte, dell'apprimento di terra in Pozzuolo nell'anno 1538 e della significazione di essi per Pietro Giacomo Toleto.* Naples: Sultzbach. [English translation in "Remarks on the nature of the soil of Naples and its neighbourhood", W. Hamilton, *Royal Society of London, Philosophical Transactions* **61**, 1–43, 1771].

Twain, Mark 1869. *The innocents abroad* (vol. II). New York: Harper.

Vezzoli. L. 1988. Island of Ischia. *Quaderni de la Ricerca Scientifica* **114**, 1–133.

Vigée-Lebrun, L. 1835–7. *Souvenirs.* Paris: H. Fournier.

Virgil [19 BC] 2003. *Aeneid* (translated by D. West) [notably Books 2 and 6]. Harmondsworth: Penguin.

Vitruvius [*c.* 25 BC] 1962. *On architecture* (2 vols; translated by F. Granger). Cambridge, Mass.: Harvard University Press.

Vogel, J. S., W. Cornell, D. E. Nelson, J. R. Southon 1990. Vesuvius–Avellino, one possible source of seventeenth-century BC climatic disturbances. *Nature* **344**, 534–7.

Walker, G. P. L. 1981. Plinian eruptions and their products. *Bulletin Volcanologique* **44**(2), 223–40.

Wallace-Hadrill, A. 1994. *Houses and society in Pompeii and Herculaneum.* Princeton: Princeton University Press.

Wideman, F. 1986. *Les effets économiques de l'éruption de 79: nouvelles données et nouvelle approche.* In Albore Livadie (1986b: 107–112).

Wilson, L., R. S. J. Sparks, G. P. L. Walker 1980. Explosive volcanic eruptions, IV: the control of magma properties and conduit geometry on eruption column behaviour. *Royal Astronomical Society, Geophysical Journal* **63**, 117–48.

Winckelmann, J. J. 1781. *Lettres familières.* In Gasparini & Musella (1991: 221).

Wohletz, K., L. Civetta, G. Orsi 1999. Thermal evolution of the Phlegraean magmatic system *Journal of Volcanology and Geothermal Research* **91**, 381–414.

Zelinsky, W. & L. A. Kosinski 1991. *The emergency evacuation of cities: a cross-national historical and geographical study.* Lanham, Maryland: Rowman & Littlefield.

Index

Page numbers in italics indicate illustrations; numbers with a "t" suffix indicate tables.